CAMBRIDGE ENGINEERING SERIES

GENERAL EDITOR

SIR JOHN BAKER, F.R.S.

THE DESIGN OF DESIGN

THE DESIGN OF DESIGN

GORDON L. GLEGG

Consulting Engineer and Lecturer in Engineering
University of Cambridge

CAMBRIDGE
AT THE UNIVERSITY PRESS
1969

Published by the Syndics of the Cambridge University Press
Bentley House, 200 Euston Road, London N.W.1
American Branch: 32 East 57th Street, New York, N.Y.10022

© Cambridge University Press 1969

Library of Congress Catalogue Card Number: 69-12432
Standard Book Number: 521 07447 9

Printed in Great Britain
at the University Printing House, Cambridge
(Brooke Crutchley, University Printer)

CONTENTS

INTRODUCTION

A designer's life is an unpredictable pattern of the challenging and the enthralling, of adventure and achievement; even learning how to be one need not necessarily be dull. It seems fashionable to glamorize the position of the scientist and to imply that no other occupation is so rewarding in human, if not material, values. I do not think that this is true, for several reasons. For instance, the engineer has the much wider horizon of possibilities. A scientist is lucky if he makes one real creative addition to human knowledge in his whole life, and many never do so.

An engineer has, by comparison, almost limitless opportunities. He can, and frequently does, create dozens of original designs and has the satisfaction of seeing them become working realities. He is a creative artist in a sense never known by the pure scientist. An engineer can make something. He creates by arranging in patterns the discoveries of science past and present, patterns designed to fit the ever more intricate world of industry. His material is profuse, his problems fascinating, and everything hinges on his personal ability. His successes and sometimes his failures become incarnate in metal. They grow up and confront him, sometimes with surprising results. A scientist can discover a new star but he cannot make one. He would have to ask an engineer to do it for him.

The purpose of this book is to suggest some guiding principles that are behind most designs and so help the young engineer on his way.

At first sight the idea of any rules or principles being superimposed on the creative mind seems more likely to hinder than to help, but this is really quite untrue in practice. Disciplined thinking focuses inspiration rather than blinkers it. The commando troops in the last war depended much more on their own initiative and resource than the normal soldier, but they also had a harder training. They were taught what others had found out before them by bitter experience. It would have been literally fatal if they had had to learn by their own mistakes. They won new ground by disciplined personal effort.

Young designers, on the other hand, often have to learn by making mistakes. Quite naturally, their instinctive reaction is to conceal such mistakes from their superiors or to invent alibis for them. The danger is that they may also conceal such mistakes from themselves. Possibly they will have to repeat the same type of error several times before looking squarely at their own failures and seeking to discover what went wrong. This way of learning will be a slow process and is harmful in at least two ways. It not only wastes time and money, but it also wastes designers. Creative ability may not always be allied to a buoyant personality that can shrug off failures. The young designer may well become depressed and decide that he will never be any good at it, and sidestep to some less demanding or routine job. Thus, the country may lose a genius, and he may miss an absorbing lifework. Negative self-help is not good enough. Positive strategy is needed and this book seeks to provide an introduction to it.

Principles of engineering design can be divided into three distinct types.

First, there are specialized techniques. Most design offices have their store of data and manufacturing techniques, often jealously guarded, without which they would not dare to attempt a new project. If you are designing a camshaft for a petrol engine, questions concerning noise, wear, cam contours and valve overlap will, to a marked extent, be decided by past experience. You are unlikely to design them from scratch, nor would you be likely to be successful if you did.

Similarly, manufacturers of, say, refrigerators or carpet machines each have their accumulated know-how which would have nothing predictably in common with the others'.

Next, we have general rules. These are broader theoretical considerations which are not confined necessarily to a single engineering mechanism. For instance, a theoretical knowledge of the out-of-balance forces in reciprocating parts would be useful in designing a petrol engine, a refrigerator, a carpet machine and many other mechanisms. But these general rules wide though their scope may be, are not of universal application, A knowledge of reciprocating forces would not help very much in designing a bridge, a turbine, or a corkscrew.

Finally, there are universal principles. These are the underlying laws which cross the frontiers of most engineering design. They are the rules behind the rules; they are not tied to any particular type of design, they concern the design of design. The purpose of this book is to define and illustrate some of them.

It should be realized at the beginning that these principles will almost certainly have two common characteristics.

The first is that when they are considered in the calm and clarity of a university or a quiet drawing office they may appear too obvious to warrant mentioning. The second is that when you are plunged into the hectic atmosphere of industry, with its inevitable pressures and tensions, you will be tempted to go to the other extreme and to dismiss them as irrelevant. That theoretical or merely abstract thinking could ever be regarded as irrelevant may seem strange until you are actually faced with a harassed managing director saying, 'We haven't time for all that theoretical stuff. What we want is some practical designing and I want it quickly.'

Such situations occur because of the increasing specialization of industrial staff. At one extreme we have the exclusively accounting mind, totally absorbed in finance and figures. At the other extreme there is the exclusively designing brain, to whom dividend stripping is as incomprehensible as differential equations are to the accountants.

Bridging the gulf between are those who have the unenviable responsibility of trying to talk the languages of both. With a few brilliant exceptions, these managers or managing directors are but half-experts in both fields. They carry the major responsibility in industry and should have support and not criticism, but it is equally true that if you are bullied by them into ignoring prudent principles of design both you and your company are going to regret it in the long run.

Even worse, a designer may be confronted with the problem of obtaining approval for a design from a committee composed exclusively of half-experts. Such committees not infrequently will approve the design provided some 'minor' alterations are made. These minor alterations may invalidate the whole principle and yet, if he starts to argue, the engineer may find the whole scheme is thrown out. On these occasions general principles of thought may seem very abstract compared with the concrete of the committee's faces. Perhaps it may help to remember then how self-evident these principles seemed when unemotionally considered, and that an ignorance of their existence does not mean that they are not still true.

1

THE DESIGN OF THE PROBLEM

Sometimes the problem is to discover what the problem is. You must aim at clarifying your objective in such a way as to make the solution easier. You are unlikely to hit a target if you cannot see it, or are looking at a different one.

There are two principles that should help in this. The first concerns the possibility of solving the problem at all. We must remember that almost all modern industrial equipment would have been considered impossible a hundred years ago. At the present furious rate of technical development today's impossibility may become an everyday commonplace in twenty years from now. One of the intriguing facts of the engineering profession is that it is not only possible but essential to do the impossible. But we must be careful to define what we mean by that word. We must distinguish between the relatively impossible and the intrinsically impossible.

The relative impossibility is something which is impossible unless something else becomes possible. It always includes or implies a clause beginning 'unless'. The intrinsic impossibility contains an imprisoned self-contradiction within itself; it is absolutely impossible.

As an undergraduate I was told that a jet turbine engine was an impossibility unless special steels were developed. It was a relative impossibility. It became possible when the steels became possible.

On the other hand, you must never waste time on trying to solve the intrinsically impossible, however tempting it may seem. Mathematics are not susceptible to negotiation, nor are the laws of thermodynamics or those of moving bodies. Try the impossible, but never the impossibly impossible. Unfortunately, in practice, the distinction may not be readily discernible.

The chairman of a company once asked me to investigate the position of a new machine on which a large capital sum had been expended. The machine was always about to go into production but never actually did. The reasons given to the chairman were contrary, depending on who was asked. Either the design was excellent but the workmanship of the components disastrous, or else the design was disastrous but the workmanship quite good. The chairman could not satisfy himself which statement was true and implied that he would not be surprised if both the design and the components were equally bad. It was up to me to find out.

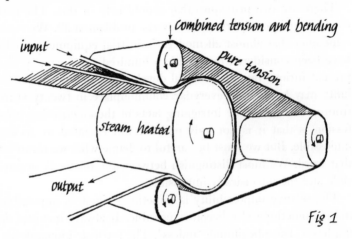

Fig 1

I went down to the factory and was shown the machine whose working principle is indicated in Fig. 1.

Layers of plastic material were bonded together under temperature and pressure by being pressed against a heated drum by a steel band. The forces within the steel band were considerable and it was located and tensioned by three guide rolls. The difficulty was that the steel band always broke after running a short time, sometimes only a few minutes. It cost hundreds of pounds to replace.

Some people were in favour of increasing the thickness of the steel band so that it would have a greater safety factor against failure in pure tension; but others thought it would make matters worse as the breakages were due to bending stresses as the band passed around the guide rolls. The basic difficulty was that the poor designer had had the roll diameters fixed for him in advance for cost reasons by a half-expert and he had done as well as he could in the circumstances. He had chosen a band thickness that gave the lowest figure for the combined bending and direct tensile stresses and now everyone was blaming him for the disasters.

I reported to the chairman that production on the machin was impossible unless

(*a*) a different type of steel band could be fitted (a link design existed but it was covered by someone else's patent and was not readily available) or

(*b*) a lower working pressure in the machine, and thus lower pure tension in the band, could be tolerated.

Unfortunately it could not. Thus a relative impossibility was turned into an intrinsic one, and there was no alternative but to abandon any further attempts at working the machine.

If the problem shows no sign of being intrinsically impossible we should bring another principle into operation, one that sounds prosaic but carries with it important safeguards.

This principle is to aim at defining the problem in figures, not in words.

Rarely is a problem initially presented to the engineer in exam-paper precision of definition. You may be lucky. You might be asked to design a change-speed gearbox and be given all the powers, ratios, and speeds required; or a structure with all the loadings and limiting dimensions; or a diesel engine with full particulars. But if to the specification of the gearbox the words 'and it must be quiet' were added, or to the structure 'it must be cheap' or to the diesel 'it must be light', you must immediately be on your guard. These words 'quiet', 'cheap' and 'light' are merely subjective and are quite useless as design factors until they have been translated into figures. In fact they are worse than useless, they are dangerous. They can easily cause misunderstandings and friction. Having produced a gearbox which you are convinced is exceptionally silent, and feeling justly pleased with it, you demonstrate it to your superior. 'It is all right, I suppose,' he says, 'but a bit noisy.'

He feels disgruntled and you feel crushed and depressed. All this could have been avoided if you could have agreed in advance on a non-subjective definition of noise level.

More important, and less obvious, is the discipline of thought that the insistence on numerical definitions carries with it.

Recently the chief engineer in a large firm was responsible for a most successful machine for the sorting of granular material into different sizes. The machine was large and powerful; too powerful for its hurriedly built foundations, for it tended to leap up and down in a rather terrifying manner. The board of the company congratulated the designer and gave him a large sum of money to build a long-term production model just like the prototype. This he did but it did not work. The machine was beautifully stable on excellent foundations but the designer tended to jump about. He soon realized that the original prototype only worked because it was allowed some freedom of movement so that various fortuitous resonances could take place. These resonances were the active working principle of the machine.

This disaster could have been avoided if the designer had defined the specification in figures, not in the words 'just like it'. He should have measured the wave-forms of the oscillations occurring in the different parts of the machine and by adding or subtracting weights or supports found how the performance of the whole depended on them. This would have given him a range of effective frequencies enabling him to build in a suitable system of mass and flexibility into the structure and not into the foundations.

I have run into difficulties myself by accepting verbal, as distinct from numerical, definitions.

I was asked to advise a company on how to reduce its labour costs by installing mechanization. There was a production department where it was obviously very necessary to do this. A patterned floor covering was being made from a plastic material. The pattern was formed by fitting together small rectangular or triangular pieces of plastic of different colours and then pressing them on to an adhesive base. These small

8

pieces were cut from sheet material about $\frac{1}{16}$ in. thick and were themselves as soft and sticky as home-made toffee. The patterns were often complicated and up to 200 pieces had to be fitted tightly together on to every square yard of base material. This was all being done by hand, with hundreds of skilful girl operators peeling off each piece from a stack and fitting it accurately amongst its neighbours. A 0·006 in. gap between each piece was the maximum allowed. For a girl to do this at a speed of about a dozen pieces a minute needed six months' training and was an exacting and tedious job. They often left after a few weeks. It was easy for the whole pattern to drift out of square or out of line and often time was spent picking the pieces up and starting again. The labour cost was terrific. Any mechanization would have to be flexible enough to tackle any one of some two hundred different patterns built up with many thousands of different piece-sizes. Some patterns repeated every 3 ft, others every 3 metres. Some were 6 ft wide, others 2 metres wide. The machine would have to run continuously, except for pattern changes, where a delay of up to an hour could be tolerated. The matter was further complicated by the necessity to deal with three different thicknesses of material on occasions.

After some thought I informed the chairman of the company that I could see no difficulty in designing a machine capable of laying and fitting together accurately up to 100 pieces per second of any combination of patterns, sizes and thicknesses that he wished, but that I would require more technical information before I could give a capital cost. To be certain that I was given all the facts, he called a conference of the heads of the technical departments and production teams. I explained to them that there were two solutions to the problem, one more complicated and expensive than the other. The simpler of the two could be used provided there was no variation from the nominal in the size of any of the rectangular pieces for any given pattern. In other words, all that would be required of the machine was the placing of the centre of each piece at the exact theoretical point in the pattern and at the correct angle. If there were any inaccuracies in the size of the rectangles or triangles being fed to the machine there would be an accumulating error and this could only be dealt with by the more complicated and expensive design of machine.

The entire conference assured me that any variation in these sizes was quite negligible and that the simpler design was all that was needed. So the simple design was built and it would not work. It would not work because the sizes of the pieces, in fact, varied appreciably due to creep or contraction after cutting. The more complicated design should have been built. It was and it worked at the promised speed. The time and expense of the first machine were an almost total loss. Foolishly, I had allowed the word 'negligible' to creep into the specification.

The interesting thing is that all the technical and production personnel quite sincerely believed that the pieces were dimensionally stable, and this sprang from the kind of folklore that you sometimes find in industry.

The company prided itself on the stability of its floor coverings and assured its customers, quite rightly, that they would experience no trouble. Some customers enjoyed seeing over the factory and it was the done thing to assure them that there was no dimensional instability at all. Everyone became indoctrinated with this idea. The girls arranging the patterns unconsciously compensated for the different sizes, which were only a few thousandths of an inch, by laying them tighter or looser, and no one was any the wiser.

I should have insisted on the word 'negligible' being translated into figures, and this I did before designing the second machine.

Another example was where a firm had the problem of weighing granular material into trays and automatically stacking them for transport afterwards. The trays were about 4 ft long, 3 ft wide and 6 in. deep, and when empty were stacked fitted into each other slightly, for their slides were sloping. When filled with material, they were best stacked by turning them horizontally through 90° each time, so that successive trays were supported on the edges of those beneath. An outside firm was engaged to design and build an automatic line which would take a pile of empty trays, fill them with a predetermined weight of granules, and then stack them ten deep ready for transport on a fork-lift truck. Some typical trays were sent to the designers as a pattern.

After some months a prototype was said to be nearly complete but there were still 'details to be finished'. Another two months went by and nothing seemed to be happening; so I was asked to go down and report on the position. A lorry full of trays had been sent down in advance so that a full working demonstration could be given.

I arrived to find a large and fairly elaborate plant ready to start up with a number of white-coated experts in attendance. One was partially hidden, as he was round at the back of the machine. The line was started up and, being of routine design in most ways, showed every sign of working, except for one point. If a tray came along that was badly bent or buckled, it needed a hurried and surreptitious push from the partially hidden assistant to ensure that a side-gripping mechanism held it firmly.

I walked around to the back and quietly asked him what would happen if he didn't prod at the critical moment. 'Practically anything' was his gloomy reply.

The whole system had been built up on the assumption that the side-gripping of the trays would be adequate and this was based on the dimensions of the typical trays. Unfortunately, they weren't typical. If the company had given the consulting firm the range of dimensions within which all their trays would lie, it would have been immediately obvious that the variations were so large that support from underneath was the only principle that could safely lift their not inconsiderable weight when full of granules.

If the words 'quiet, light, cheap, just like it, negligible' and 'typical' are fairly obviously unreliable when isolated from the camouflage of local colour, the words 'instantaneous' and 'simultaneous' seem to imply some technical exactness and are all the more dangerous for doing so. They are never technically exact. No two events ever literally happen 'simultaneously' nor could we design any machine which made them do so. Everything occupies a finite time.

I found an example of this in a drawing I was asked to approve. It showed a conveyor belt that travelled at a number of different fixed speeds and carried upon it articles that had to be processed at the right points in space above the conveyor. A camshaft was driven from the conveyor drive and the processing operated by compressed air cylinders, controlled through air relay valves operated by this camshaft. When the conveyor went faster so did the camshaft. This was the thing to do but unfortunately the draughtsman had overlooked the fact that the air in the pipes still went at the same speed, i.e. the lag between the relay valves being operated and the process cylinders operating remained constant. As a result the whole thing would need retiming each time the speed was altered. I wondered why the draughtsman had assumed that air-operated valves worked instantaneously. He said that as a hobby he played the organ and could detect no pause between the depression of the keys and the organ responding. From a musical point of view the word 'instantaneous' was in practice correct; it was the moving of this word sideways into the engineering world that was his unconscious error.

Probably it is in process engineering that this principle of numerical definition is most necessary and also most easily overlooked. If you are concerned in making the production of some article quicker, better or cheaper, you will naturally study how it is being done at the moment. It may seem paradoxical but the man least likely to help you is the operator in charge of the machine. The more 'down-to-earth' and 'practical' he considers himself the more 'unpractical' and 'up-in-the-air' he is likely to be from your point of view. He is too near the machine to know why it works. He only knows how it works. He will have learnt what handwheels to turn or what controls to adjust to keep the line running efficiently, and will be able to use that knowledge most skilfully and responsibly. All this he will have learnt from experience and then memorized, and to help in this he will, probably unconsciously, have formed a picture in his mind of the working principle of the machine. This picture may be nothing more than a mental coat-hanger on which the operating facts can be neatly hung up and found quickly. When you ask him what goes on in the machine, he will describe this home-made mental picture and demonstrate its accuracy by showing you how well the operating facts fit into it. He will be very sincere and convincing and dangerously misleading. You must find out the facts yourself.

Once I was asked to design a faster and more effective version of a machine that rolled out plastic sheet. The patterned appearance of the surface of the sheet was a most important sales asset and had to be maintained or improved. The process

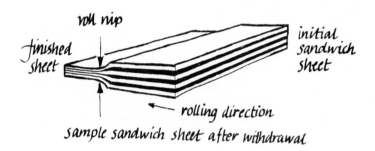

sample sandwich sheet after withdrawal

was a complicated one and was completed by passing the materials through the nip of a pair of heavy rolls. These rolls had greatly different speeds and temperatures, and the surface effect was known to depend on the adjustment of these conditions. Everyone, especially the man in charge, described most vividly how the material rotated in the nip of the rolls and how this was the secret of maintaining the appearance. However convincing this sounded, I felt I had to have confirmation. I arranged for the machine to be fed with a sheet of material that was specially prepared in the form of a multi-layer sandwich of very thin black and white sheets. This composite sheet was fed into the machine in the normal manner but after running a few feet through the nip, the machine was stopped. The electrical connections were then altered so that the machine started up in reverse, and regurgitated the sample sheet in the form of a wedge. When this was sectioned longitudinally the black-and-white pattern gave a visible and measurable picture of what was happening in the nip. It was immediately obvious that, despite all the handwaving and explaining that had been done, the flow was a strictly laminated one. We had to avoid and not maintain a turbulent one.

An extension of this principle of objectivity is the use of a three-dimensional model. Often configurations are better than figures, for nothing can be more objective than an object. If your problem concerns the squeezing of a design into a confined space or assembling an involved unit, a full-size or scale model may be much the best way of defining the boundaries and discovering if everything will fit in.

Sometimes you may have to go further and construct an apparatus to test the problem. If you are designing a new car you may well make a prototype and bash it around the roads of North Africa. This will not only tell you something about the car but it will test the roads too. To the question 'What are the roads like out there?' you may well reply, 'Look what they did to my steering-linkage.' By reproducing the same damage in a test machine you will be able to translate the road conditions into figures and forces.

The contrast between the strictly determinate and indeterminate I first met as the two halves of the same problem in the early days of motor racing. Between the wars, the Brooklands track at Weybridge was the mecca of much motor racing and, in addition to its level oval track, there was a test hill. This hill was constructed with great precision. Dead straight and with a smooth concrete surface, it started with a gradient of exactly one-in-eight, and maintained this for some distance. This was followed by a section of one-in-five, and finally a longer one of one-in-four. The average gradient was slightly over one-in-five.

There was some rivalry over who could ascend it in the shortest time from a standing start and runs were electrically timed to one-hundredth of a second for record attempts. As speeds increased it was found that the fairly sudden transition from one-in-four to level at the top of the hill caused the cars to leap into the air. The existing record-holder found it essential to brake hard well before the summit to avoid being airborne too long.

I thought it would be nice to capture the record. Being an impecunious undergraduate at the time, I had neither the facilities nor the finance to build a special engine and so I would have about half the power of my rivals.

On the other hand I had devised a four-wheel-drive transmission which might help to lessen the disadvantage and, if I went flat out over the top, I felt the record could be broken.

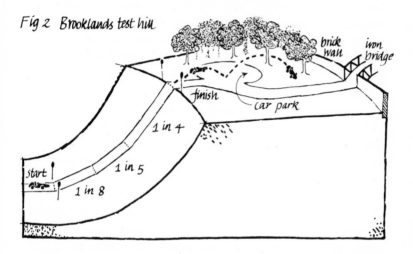

Fig 2 Brooklands test hill

The first half of the problem was very precisely laid out. The exact gradients and lengths were known. From these I could calculate, quite accurately, the gear ratios needed and also the speed at which I would be going over the summit. I would then take off into the air and gravity would take over (Fig. 2).

After the top of the hill the road went straight for a short distance and then curved off to the right, leaving a gravelled car park on the left. Behind this again was a small wood. I calculated that a car going flat out over the brow of the hill at record-breaking speed would not come down again until it landed in the middle of the car park. There the precision of the problem ended. I could only guess what would happen next. The car would probably take off again in a series of bounds and I would be dodging trees in the wood before I could put the brakes hard on and steer at the same time.

This sudden transition from the known to the unknown was a major problem in designing the suspension. The hill itself was smooth and predictable but the loose-surfaced car park was quite the reverse. Moreover, the car might very well be pointing in one direction and going in another when it started to leap, and would preserve this crab-like attitude in the air. When it landed, possibly on two wheels, possibly on one, the sideways forces would combine with the impact ones. To make the suspension extremely flexible and so allow a large movement might have swamped the problem, but this was not possible. The four-wheel-drive transmission was to use fabric-disc in-board universal joints, for they were both light and cheap, and a large angular deflection would tear them. As I was depending on a transmission brake, again for lightness, a collapsed universal would mean no brakes and thus add a further complication to a situation that might be quite complicated already. One, quite literally, jumped from the predictable to the unpredictable.

There was only one solution and that was to do full-scale tests. The car was built with a suspension intentionally over-flexible. Means for measuring deflection were fitted and a series of runs of gradually increasing speeds gave figures that, when translated into equivalent static loads and plotted, allowed a reasonable forecast to be made of what would occur when going at full speed. The suspension was then designed to suit.

Of course there are always things one overlooks: such as mud flying up and blacking out one's goggles, while attempting to dodge the more substantial trees; three-dimensional determination often gives unexpected and useful fringe information in this way. It throws up unanticipated complications.

To sum up, the engineer's job is to design an objective solution to an objective problem. He should not be expected to solve a purely abstract one and would be wise not to attempt it. He may be presented with a general problem and have some general ideas on how to tackle it, but he cannot design the particular without boundary particulars.

This chapter may be summarized by saying:

(1) Beware of intrinsic impossibilities
(2) Beware of pseudo-technical words
(3) Define problems in figures or configurations

2

THE DESIGN OF THE DESIGNER

Creative design is an essentially personal achievement. There is nothing automatic about it.

If you do not understand a foreign language you can painstakingly look up everything in a dictionary. It takes longer but it works in the end. If, as a designer, you become puzzled over a problem there is no routine way out. You cannot always find the solution in books. You may have to find it in yourself. The guiding principles we will discuss are not to help the design but the designer: principles can't use pencils.

Whether he is conscious of it or not, the mind of the designer has three realms of activity. We will discuss later whether it is wise to departmentalize thinking in this way, but for classifying overall principles this distinction is useful. For this purpose we will consider that the creative mind can be subdivided into the inventive, the artistic, and the logical or rational. These realms of thought will become clearer when we consider detailed design problems and at this stage it is only necessary to define them briefly.

The inventive

Everyone wants to be an inventor. We dream of creating some wonderful new machine and living in luxury ever afterwards. There is no reason why this dream should not come true but the process is likely to involve much more hard work than was mentioned in the dream. Rarely do inventions fall like a bolt from the blue; they have to be conjured up from the conscious and subconscious mind. You cannot command your mind to invent something, but you can encourage it. The best way to do this is by saturating your mind with all the elements of the problem. Study everything you can; try to find the feel of the job.

When I was a small boy, a neighbour of ours called Mrs Roe

said that her husband always wanted to spend his holidays in the same way. He sat on the top of a cliff at the seaside and watched the seagulls. He studied them for days on end. We all thought this to be highly eccentric, but it was really highly sensible. It enabled him to be a pioneer in aircraft design. He was A. V. Roe of the Avro aeroplane.

The secret of inventiveness is to fill the mind and the imagination with the context of the problem and then relax and think of something else for a change. Perhaps you could read a book, play a game, or climb a mountain and thus release mental energy which your subconscious can use to work on the problem. If you are lucky, this subconcious will hand up into your conscious mind, your imagination, a picture of what the solution might be. It will probably come in a flash, almost certainly when you are not expecting it. This is true of all creative thinking whether in engineering or not.

History tells us that in a selection of fifteen creative artists in various fields from music to mathematics, their key inspiration came suddenly and unexpectedly and never when they were working at it. This is what they were doing at the time:

Half asleep in bed	4
Out walking or riding	3
Travelling	3
In church	2
At a state dinner	1
Sitting in front of fire	2

Concentration and then relaxation is the common pattern behind most creative thinking.

It is also important to realize that our subconscious minds will hand up their suggestions in the form of symbols or pictures. The subconscious has no vocabulary. To encourage communication between the conscious and subconscious, we should practise their only common language, which is in three-dimensional pictures. That is why all engineers should learn to do three-dimensional sketches.

19

The artistic

The next subdivision of the designer's mind is the artistic and it is much the most difficult of the three to define. The sense of the artistry of engineering is invaluable but cannot be formally stated. A machine may look artistic in the normal meaning of that word without being good engineering. A bridge may look nice but fall down. An efficient high-voltage insulator is often ornamental to look at, but if you designed it merely for perfection of form it would not necessarily be a good insulator. The artistry of engineering is essentially a matter of style and this is always a problem to put into words. We find the same difficulty in talking about style in music, literature or art. The definition is often only appreciated after the style has been recognized. Writing about the world of science, Sir Arthur Eddington observed, 'We sometimes have convictions which we cherish but cannot justify; we are influenced by some innate sense of the fitness of things.' Perhaps this 'sense of the fitness of things' is the nearest we can get to a positive definition of engineering artistry too.

Surely a self-aligning roller bearing commends itself as good in principle! And is there not a simplicity of style in the design of a squirrel cage motor? I sometimes wonder if such a humble thing as an umbrella is not a remarkable example of structural engineering. Very few structures can be erected or pulled down so quickly.

We must be careful to distinguish between a good principle of working and good workmanship in applying a principle. Probably most of us will remember the first time we prised off the back of a watch with a pocket knife. We were immediately struck by the fascinating delicacy and precision of the diminutive components, but this of itself said nothing about the essential style of the working principle. Did we go on past this perfection of workmanship and appreciate the economy of style in the principle? The precise metering of energy by a small spring from a bigger one is an engineering joy. The total energy that would only sustain a child's top for less than a minute, despite its low friction bearing, will run a watch for a day and, in total contrast to the top, at an exceedingly uniform angular velocity.

For transporting its own weight over rough ground, a child's hoop is artistically good while a motor bicycle and sidecar, even though designed to a high state of engineering sophistication, is clumsy in principle.

This distinction is important, for it is nearly always true that some new breakthrough in engineering design may initially appear less mechanically sophisticated than the highly developed traditional one it will soon replace. Style will always win in the end.

The rational

The third realm of activity is the rational, which represents disciplined thinking applied over the entire field of design from theoretical analysis to economic realities. The inventive and the artistic, the inspirational and the intuitive must all be impartially scrutinized. The rational must hold the power of veto over them all.

The reason for this is that all machines and structures are inherently rational; they always work exactly to theory. They obey with 100 per cent accuracy the circumstances of their construction and environment. A machine never makes a mistake here, it could pass any exam in its own subject. Theory is practice. Thus, paradoxically, the more expert we become in theoretical analysis, the more we approach the actual working of a machine and so the more practical we become. A machine does not guess. We must reduce our guesswork to a minimum, and so safeguard ourselves against emotional or illogical decisions.

For instance, the artistic style of a roller bearing may tempt us to use it indiscriminately, but in some circumstances it is useless. Logic reminds us that the higher the rotational speed, the greater is the centrifugal self-loading of the rollers themselves. So that the bearing will not overload itself by internal forces it must be made smaller and smaller as its speed increases. At 80,000 rev/min a quarter-inch-diameter bearing uses up nearly all its strength in avoiding flying to bits. To avoid self-destruction at much over 100,000 rev/min the bearing has to be so small that it almost ceases to exist and therefore is not a very useful component. Neither our sense of style nor inventiveness would, of itself, erect the necessary frontiers. The logical is the watchdog of design.

Vital as this is, we must beware of going to the other extreme and regarding all design as a strictly logical exercise. It is no substitute for the inventive or the artistic. Logic may decide between alternatives but cannot be relied upon to initiate them.

For some years I employed a consulting engineer who was a brilliant designer of components. I greatly admired the strict logical analysis that he applied to his problems. He wrote copious notes and one could follow his reasoning in detail. Unfortunately, he had no sense of style. His designs for complete machines or processes were often impracticable and the remarkable thing was that they were arrived at by painstaking reasoning. Every step appeared, from his notes, to be a logical development from the previous one; but one could see, with consternation, the gradual drift towards the unrealistic. Aiming for simplicity he would achieve gawkiness, for sophistication the hopelessly involved. Logic is not enough; a sense of the fitness of things is needed too. Often the sudden impact of seeing a sound design produces an instantaneous response. Afterwards we can support our approval logically but we know in advance what the verdict will be.

There was a large warehouse where many thousands of rolls of floor covering were stored. Unfortunately there was no doubt that they were being steadily stolen. Special vertically sliding steel doors were fitted leaving a gap of less than a quarter of an inch anywhere, and it was thought that it would need a clever man to engineer rolls of linoleum 6 ft long and 18 in. in diameter through any of the cracks. But it was a clever man who was doing the stealing and he went on doing it. Most nights, just before leaving the warehouse and after everything was locked up, he would push over a roll of linoleum and roll it along the floor until it rested against the bottom of a door. He would then unroll the linoleum slightly and tuck the leading edge under the door, through the quarter-inch gap between it and the rough concrete floor. Late that night he would return to the street outside the warehouse and pull on the edge of the linoleum he had left slightly protruding under the door. He would go on pulling, and winding up as he did so, until he had a complete roll of linoleum on his side of the door, and then off he went with it. Such criminal exploits have nothing to recommend them except a simplicity of style. The thief might have become a good process engineer.

Returning to the realm of the logical, we must extend its scope beyond that of being merely a mathematical referee to include the economic. It would be nice if engineers didn't have to worry about money; very nice indeed. Unfortunately the real position is exactly the reverse. In general, the purpose of engineering design is to spend money in order to make it—to make a great deal of it if possible—and, most important of all, to raise the standard of living for the community. Often we would like to construct a machine just for the fun of seeing an ingenious principle at work, and sometimes we are tempted to think that the more engineeringly subtle a machine is the more money it is likely to make. The logical part of our thinking is the great enemy of all such wishful thinking and that is sometimes why we are reluctant to use it. I knew of a firm whose managing director devised what he thought to be a marvellous labour-saving machine. His firm employed a large number of men who, working in pairs, assembled large sheets of material in mesh form. The managing director ordered the machines to be built and these did the assembling nearly as quickly as the men. Unfortunately the finished article needed careful handling and no facilities for mechanizing this had been thought out. The two operators therefore had to be retained to lift the product off the assembly table. The design had used capital, slowed output and left labour costs slightly higher than they were before.

At this stage we have probably said enough about the three categories of engineering thinking to enable us to follow how the rules for design are classified in later chapters, when the definitions will in any case be clarified by more detailed illustrations. However, something further needs to be said about the design of a designer to avoid giving the impression that all competent engineers must be equally at home in all branches of design. It is probably rather rare to have a design mind equally at home everywhere; and from a career point of view facility in one realm only is quite enough. You can have unbounded opportunities provided you know what you are good at and go where you have an opportunity to exploit it.

Most larger engineering firms have three internal departments dealing with new designs. Although not always called by the same names, these are 'the design or project department', 'the development or prototype department' and 'the production department'.

The design department is responsible for new ideas; development clothes them in mechanisms and then, if the prototype works well enough, production takes over to refine and streamline. It follows that these three categories of industrial organization broadly correspond to the three categories of engineering thinking. So, if you are an inventor by nature, head for the research and design departments; if artistry of style fascinates you, go into development; and if you have a logical or mathematical mind, they will need you in the production of the final machine.

If you have an inventive turn of mind you may have discovered it quite early in your life. Did you ever play with Meccano or electronic kits? And were you interested in making your own designs rather than those shown in the book? Perhaps you rarely tired of making things but often tired of them after they were made. If any of this means anything to you there is probably an inventive streak in your mind. If it doesn't, it may only mean that you have not had the right opportunities. No one need eliminate himself until he has tried and tried again.

A sense of style is something that tends to develop over the years. It is rare for anyone who is proficient in the other two spheres to have no feeling for it. The consulting engineer I have already mentioned is the only exception I have known.

Easiest of all to recognize is the mathematical and logical end of the engineer's spectrum. If you can do maths, engineering is waiting for you and there is the exciting new world of computer-aided design that beckons you further on still.

There is one overriding qualification: you will never be a good engineer if you are frightened by hard work or responsibility.

3

THE DESIGN OF DESIGN
THE INVENTIVE

The inventive in design is obviously a matter of individual enterprise. Some suggestions for an approach to it have already been made. Now we must consider what general principles may be helpful. Concentration and relaxation are the essential preliminaries, but there is no logical order for the subsequent signposts, except, perhaps, for the first one which is: 'Do not be conditioned by tradition.'

This does not mean that you must never consciously follow traditional design; it means you must beware of subconsciously doing so. The danger is that one's subconscious may not hand up an original picture but unhook a copy of someone else's masterpiece and we do not recognize it for what it is. Examples are not hard to find.

A great deal of engineering design in the past, and even some in the present, would have been quite different and much better if its originators had never seen a horse and cart or a railway train. I am not belittling the design of either; in fact the danger arises precisely from their excellence of design which has given them a traditional authority which we may uncritically accept. Do not despise the design of a horse-drawn cart or waggon. Its artistic style is good. 'California or bust', cried the early pioneers; and, flying over the amazing mountains and deserts that they traversed, one realizes what a triumph of simple engineering it was that enabled them to reach the Pacific.

The mistake was that when motor cars were invented, people subconsciously began designing their chassis like carts, thereby overlooking the elementary fact that while on the former the wheels pushed the chassis, in the latter the chassis pulled the wheels. They devised a 'horseless carriage' and forgot that the horse was missing.

To say that the engine of a car traditionally was always in

the front because the horse also was is probably an over-simplification. The rough roads of those days would have seemed to demand large spring deflections, and the further the transmission was from the rear axle, the less angular movement would be required from the universal joints. Useful, too, was a starting handle fastening directly on to the end of the crankshaft without the doubtful benefit of bevels. Also, the gymnastics associated with starting the engine could be performed with more abandon at the front, rather than at the side of the car, especially if the garage was narrow or the passing traffic dangerous.

But when we look at the chassis and suspension design it is quite another story. For over a quarter of a century the cart tradition predominated. In a cart, where the forces are largely all pull or push, the basic frame design is that of straight members rigidly mounted at right angles. The early car designers, disregarding the fact that a car's engine and brakes produce an embarrassing amount of twist, carried on with the cart complex.

The total length of the frame members in a typical chassis of those days, and even until the 1930s, came to between four and five times the length of the wheelbase. The first mass-produced car in this country to break with this tradition was the Austin Seven.

My brother and I bought one for £5 when we were at school. Bringing it home and attempting to discover the maximum speed, my brother did a double-forward somersault in it. This remarkable manœuvre he accomplished without harm to himself. It was a different matter with the car. The result suggested that this was easily the quickest way to detach the bodywork from the chassis. Those bits which were not scattered over a wide area of Wimbledon Common came off quite easily and I saw with exhilaration for the first time a chassis frame that was nothing like a cart. Someone had been thinking, not just copying. He had joined up the points exerting or receiving loads with the shortest possible lines. The suspension was achieved by making the ends of the frame members flexible. As a result the total frame length was reduced from 4–5 down to $1\frac{2}{3}$ the length of the wheelbase. Moreover, instead of the wretched springs having to withstand all the driving and braking torques and forces, there were separate members solely designed for this purpose. The engine was used to stiffen the chassis and the side members passed under the centre of the seats; there was engineering artistry at work, the cart complex had vanished at last (Fig. 3).

The traditional layout of the steering gear is another example. It took nearly fifty years for designers to realize that pneumatic

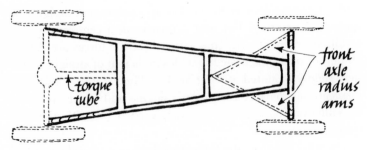

Fig 3   Austin 7 chassis frame

and not iron tyres were being used on motor cars. With iron tyres the wheel always wants to go in the direction that the rim is pointing. With rubber tyres it almost never does. If there is any force at right angles to the direction of rotation in a pneumatic tyre the part in contact with the road distorts by moving bodily sideways. To counteract this and maintain the original rolling direction, the wheel must be turned through an angle to compensate for it. Rather misleadingly this is called the 'slip angle' when it is really a distortion angle, and may be as great as 10 per cent. There is no slip angle with rigid iron tyres protecting the wooden cartwheels. The Ackerman linkage was first patented for use on carts to ensure that the inside wheel turned through a greater angle than the outside one on corners, and this was just what was needed with solid tyres. Unfortunately, generations of cars uncritically adopted it too. Now almost none do, for it is realized that the slip angle has stood the basic assumption of the Ackerman principle on its head, by transforming the geometry of the problem (Fig. 4).

Fig 4 Ackerman steering linkage

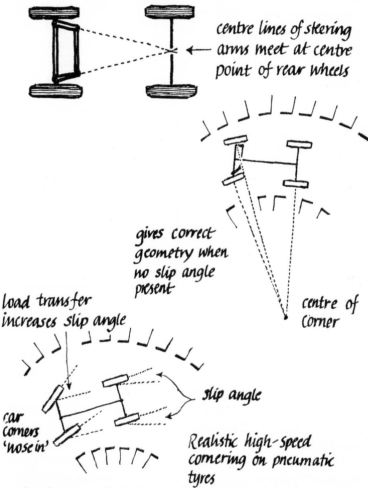

centre lines of steering arms meet at centre point of rear wheels

gives correct geometry when no slip angle present

load transfer increases slip angle

centre of corner

slip angle

car corners 'nose in'

Realistic high-speed cornering on pneumatic tyres

Another example is the way that designers' minds, as well as railway engines, have run along railway lines. The subconscious conditioning of engineering by locomotive design may have started with the model railway in the nursery. No one can tell when it will finish. When someone wants to transport anything on rails it is generally railway lines that he thinks about first. These have considerable limitations and we run into impossibly difficult situations if we try to go round right-angled bends with them; any bends are a nuisance. Quite recently a new design of rails for conveying has been patented; it is simply a railway wheel and rail turned through 90° (Fig. 5). The remarkable thing is that the novelty of the design is sufficient to be patentable. The traditional respect for the design of railways seems to have erected a buffer to thoughts of new developments in some

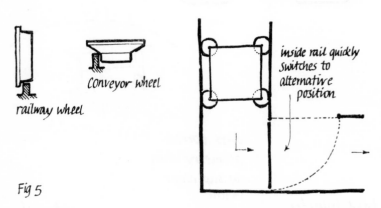

railway wheel

Conveyor wheel

inside rail quickly switches to alternative position

Fig 5

instances. 'Streamlining is not worth it,' we used to be told, 'otherwise the railways would do it more.'

After carefully analysing the reasons, by all means follow traditional design if you feel it to be wise, but beware of being subconsciously mesmerized into doing so.

Turning from the negative to the positive, there are three principles that are important and I will give them in the order that I learnt to appreciate them.

The first is that often one must complicate a design so that the overall result is simpler. If a design is becoming unbearably involved it is often because we are making one part or component too simple.

The first problem I ever had to face in a drawing office was that of designing a drilling head for the manufacture of acoustic boards. In those days these board were constructed from a fibrous material most difficult to drill because pieces of the fibre clogged up anything they could. Holes of $\frac{1}{8}$ in. diameter spaced at $\frac{3}{4}$ in. in each direction had to be drilled in successive slabs of material 12 in. × 12 in. and so over 200 drills were in operation at the same time. Inevitably some drills bent and broke and so facilities for quick replacement were essential, and this implied the ability to remove broken drills without delay.

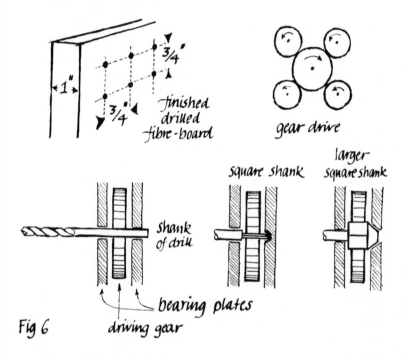

gear drive

square shank

larger square shank

shank of drill

bearing plates

Fig 6   driving gear

finished drilled fibre-board

Faced with this problem you might possibly follow the same pattern of thinking as I did. First you would say to yourself that to spin all these drills round in such close proximity, the simplest way would be to gear them all together. To keep them all turning round the same way a common central driving gear for each set of four would be needed (Fig. 6). Obviously the drills are not strong enough to withstand any overhang, so we must have bearings close up on each side of the gears. Holes in some bearing material would do. How do we fasten the gears on to the drills? With a set-screw? You immediately reject this idea, which possibly arose from your subconscious memories of playing with Meccano, as obviously the set-screws would come out with the vibration. This was before the days of special self-locking screws. In any case it would be essential to grind a flat on the shank of the drill otherwise the screw would not grip it well enough. But that would make a local stress and the drills would break there. Rejecting the Meccano idea, you next think of a square ending to the drill, and a similar hole in the gear, and this would transmit the drive suitably. Unfortunately it would not take the thrust of the drill. You think you can overcome that by making the drill with a conical end to form a thrust bearing as well as a locating bearing in the inner plate.

You feel you are getting on nicely until you suddenly realize that the square section is weaker than the round one and it will break off. Well, you will make the square bigger so as to build up to the original strength. You sketch this down and look at it. Then you realize, with horror, that you can't get the drill out to replace it; the square section is too big to get through the round bearing hole. You can't think what to do next and so you turn your attention to the gears. Each quartet of gears has a centre driving gear, which means over 300 of them in all, and these will have to be driven by others and so on until we have reduced the movement to the single input shaft which is specified. How about lubrication? It will have to be oil at that tooth speed. Will it leak? Yes, for we cannot complicate the drill assembly by having oil seals too, especially as they would not last long with that fibre dust in them. You begin to view your drawing board with distaste. Hundreds of drills which must be quickly replaceable and can't be; so many gears that you have lost count. If a tooth comes off one of them it will do disastrous damage. And it will take days to assemble. And ruin the fibre boards by leaking oil on to them. It is exasperating. You rub it all out and gaze dumbly at the now slightly grubby sheet of paper in front of you.

At this moment a senior colleague comes in and says brightly,

'Got that drilling head mapped out yet?' You glower at him and mutter something inaudible. Afterwards you hope it was inaudible. You feel that there must be some simpler solution. There is one. You could drive all the drills by one input shaft, using a total of three gears only and have quick replacement too. And the secret of discovering how to do it is to realize that all the complications you are in are due to your having made one part too simple. There is another way of rotating a shaft than by a gear. Put a handle on it (Fig. 7). A drill with a cranked handle is more complicated and expensive than the orthodox

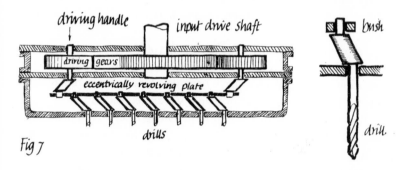

Fig 7

straight shank but its advantages are overwhelming. You can turn the handle by putting a bush on it and turning that. To stop the handles hitting each other you must arrange them to keep in step; easy enough: fasten all the bushes together by making them holes in one plate. How can we make this plate describe the proper path? Drive it by a couple of handles put in the other way round. As they must rotate the same way, put a gear between them. This gear can be on the input drive shaft.

To replace a drill, all you need to do is to remove the outer bearing plate, pull out the broken drill backwards, put in another, and replace the plate.

All this bursts into your mind in a flash, much quicker than it takes to describe, and you feel inordinately pleased with yourself. Designing is fun after all. But you must submit it to the critical eyes of your artistic and logical faculties. Their views are vital. The artistic approves the economy of style and adds that the reciprocal use of handles is a pleasing touch. The logical is more argumentative: it nearly always is. 'It is all out of balance as you have shown it', it remarks. You think for a moment and reply, 'We can deal with that by putting weights on the two driving handles.' 'These driving handles have too much overhang', it objects. 'All right, make them eccentrics and put in some ball bearings', you say. 'What about the oil leaks?' is the next question. 'We can use grease now', you reply.

The logical says, 'Well, I approve, but I am going to watch you carefully over the details, especially re-registering that front plate after it has been removed. It ought to slide in and out on guides.'

By complicating the simplest part, you have simplified the complicated part. Of course, since then, more sophisticated solutions to this problem have been developed, but it remains a useful illustration of the principle we are considering.

Another strategy of thought in inventive design is to make an ally of one's material. Often, especially in the past, many machines appear to be having a running fight with it. Avoid smashing and grabbing if you can. Peace is better than war.

For example, the traditional method of driving piles into the ground is a horrible sight and sound. A large weight bashes remorselessly on the end of a steel tube. The noise drives people mad, the earth shakes and the near-by buildings rattle, and the energy flies off uselessly in all directions. No one loves a pile-driver.

Now when a steel pile is suddenly hit on its top, the bottom does not simultaneously go further into the ground. The force from the hammer is transmitted as a shock wave which travels down the length of the pile, finally giving the bottom end a kick. But more than this happens. A steel pile behaves like a long helical spring and vibrates longitudinally at its resonant frequency, until the side damping effect of the ground dissipates the residual energy, which it does very quickly.

The problem of making the tube an ally and not an enemy can be solved by utilizing this resonant frequency. If you apply

34

a force that pushes and pulls on the top of the pile with the same frequency as the resonant one, you can silently feed in energy in a form which the pile can transmit to its bottom end. This has useful side effects as the vibration fluidizes the earth surrounding the pile and greatly reduces the frictional drag. The energy is tuned, focused and economized.

Another example can be found in a problem that has already been outlined concerning the placing of pieces of plastic in a pattern on a background sheet. The material was 'as soft and sticky as home-made toffee'. Perhaps it appears surprising that this fact was not referred to as an example of words being used instead of figures. The reason for the omission was that any reliable mechanization of the process must essentially depend on eliminating the stickiness altogether, otherwise we would have a fight on our hands. It was well known that the higher the temperature the greater became the adhesive properties of the plastic. The obvious thing was to reverse this process and freeze the material. This was done and the stickiness vanished. It transformed the situation in principle. The pieces were now as crisp as postcards. They could be piled up in a stack without sticking to each other. This fact became the basis of the new machine. Instead of a large number of girls laboriously assembling each piece of the jigsaw by hand, a girl would assemble a 3 ft length of the pattern out of a pile of a hundred or more pieces on top of each other. This could be fed into a machine fitted with a suction box where some 5,000 rubber suckers on the lower surface would pick up a single layer of the pattern at a time and deposit it on the continuously moving backing material. When one pile was exhausted another pre-prepared stack was taken out of the refrigerator room and automatically put into position without the machine having to slow down. The accumulated error due to shrink or creep in individual pieces was automatically compensated for by making the suction head lay the pattern plus or minus a fraction of an inch from the nominal 3 ft repeat when this was necessary. This principle is shown diagrammatically in Fig. 8 and will be explained in more detail later.

Some years ago I was confronted with a three-part mechanization problem where the solution to each part was made almost ludicrously simple by co-operating with the materials. Steel reinforcement for concrete was being manufactured by strain-hardening square-section mild steel bars through twisting them. The resulting spiral form also had the advantage of gripping the concrete better.

For small sections the mild steel was delivered to the factory in coils of about 3 ft in diameter. The steel was then pulled

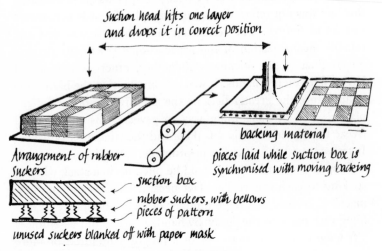

Suction head lifts one layer
and drops it in correct position

Arrangement of rubber suckers

backing material

pieces laid while suction box is synchronised with moving backing

suction box

rubber suckers, with bellows

pieces of pattern

unused suckers blanked off with paper mask

Fig 8   Diagrammatic layout of laying machine

off these coils and partially straightened by passing it through a sequence of grooved wheels, but some kinks were still left in. This meant that it was impossible to measure accurately the length of steel that had been drawn out from the coil. After twisting it became easy to measure, for it had become dead straight. In the existing set-up, the twisting was done separately. First a length of steel was drawn through the straightening rolls and an amount cut off which the operator hoped, after twisting, would be exactly 64 ft long. Rarely did he guess right. To be on the safe side he always had to overestimate the length and thus there was always a small piece that was wasted. The first problem was therefore to discover a means whereby the material could be cut off so that, after twisting, it became exactly 64 ft long.

The solution was to install a continuous twisting mechanism into which the steel from the coils was fed through a hollow chuck and then automatically measuring the length and cutting it off after it had been twisted. In other words, merely reverse the order. Twist first, cut the length afterwards. Make friends with it (Fig. 9).

The second problem was the degree of twist to be applied to the steel. The mild steel was far from uniform in its tensile

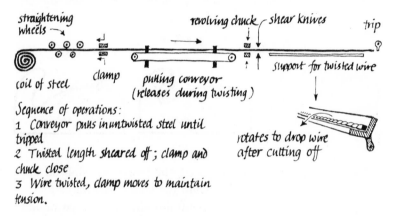

Sequence of operations:
1 Conveyor pulls in untwisted steel until tripped
2 Twisted length sheared off; clamp and chuck close
3 Wire twisted, clamp moves to maintain tension.

Fig 9 Cutting and twisting machine

properties. Both the yield and the ductility varied and tended to do so inversely with each other. The strain-hardening increased the value of the yield and reduced that of the ductility, but did nothing to reduce their initial basic variations, which were handed on to the twisted finished product.

As the code of practice asked for certain minimum figures for both yield and ductility, a compromise amount of strain-hardening, which was the method used at that time, meant that some of the reinforcement tended to be too weak and some tended to be too brittle.

Here again the material itself came to our aid. The torque

reaction to the twisting operation was related to the yield point of the steel. The machine was therefore arranged to stop twisting automatically when a predetermined torque figure was reached. By this means, the higher the initial yield the less hardening was done, which was exactly what was wanted. Initial variations were effectively smoothed out, and its energy absorption as reinforcement in concrete could be nearly doubled in extreme cases.

The third associated problem concerned the twisting of the thicker bars, a process which was done in a long lathe-type machine. Bars an inch square or more and 50 ft long were held at one end and quickly rotated at the other. When the desired

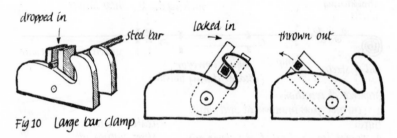

dropped in   *steel bar*   *locked in*   *thrown out*

Fig 10   Large bar clamp

degree of twist had been obtained, the power was shut off but the elastic energy in the bar spun the chuck backwards for an appreciable time, overran the neutral position, and then wound itself the other way, and repeated this until all the energy had been removed. This took time and was potentially dangerous if the operator released the fixed end of the bar too soon, for it would spin out and jump around. The solution was to make an ally of the residual energy of the bar and employ it to unclasp automatically the end of the bar immediately it had returned to the unstressed state and before overrunning. The design of this is shown in Fig. 10.

These examples, which have been chosen for their brevity rather than their subtlety, should be sufficient to indicate the principle that it is better to make friends than enemies in engineering as well as in other fields.

The next underlying principle of inventive thought is that

of dividing up and tidying up. Sometimes an engineer is faced with a novel problem where there is no precedent to lead or mislead him. He may be appalled by the overall complexity of what he is asked to do. Often this numbing impact may be overcome by subdividing the problem into a number of little problems and then putting their solutions together, like a row of dominoes. By treating each problem as if it were the only problem, he can focus all his attention on it. This elaboration of the problem has one great danger. You may end up with a design that is unnecessarily expensive. If you elaborate you

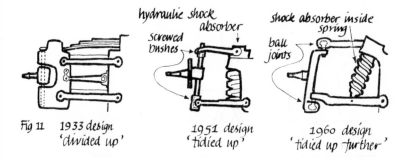

Fig 11   1933 design   1951 design   1960 design
         'divided up'   'tidied up'   'tidied up further'

must then refine; if you divide up you must tidy up. You might never guess that some delightfully simple designs were refinements of elaborate ones. They were not invented directly.

Perhaps the evolution of the suspension systems of cars may give a simple introduction to this principle.

For some thirty years the cart complex reigned almost supreme, with Morgan and Lancia notable exceptions, and then everyone woke up to the fact that as bumps on the road often occurred on one side or the other it was a pity to arrange things so that both sides of the car were affected by any one of them. So everyone decided to have individual suspensions, designed so that individual bumps could be absorbed. Fig. 11 shows how the early designs largely depended upon individual components to deal with individual forces. Nowadays we have much tidier designs where one component deals with a number of forces simultaneously. This is not an ideal example because of the

39

long gap that occurred between dividing the problem up and tidying it up, which ideally should be done at the same time.

Imagine that you are faced with a process problem where a granular material must be spread uniformly on to a conveyor. The material itself has a range of particle densities that are unpredictable and constantly varying, while its bulk density, depending on how much it is squeezed together, approaches a $2:1$ range, you are told. Combined, they may well vary by $2\frac{1}{2}:1$. This material must be spread evenly on to a moving conveyor and maintain a constant weight per foot run, to within limits of

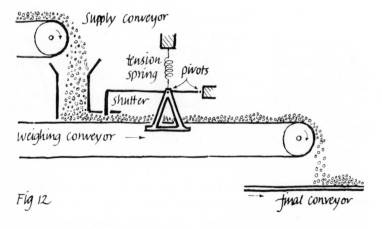

Fig 12

± 5 per cent. You are not being asked to design a process from scratch, but to make an existing machine work more accurately.

You are taken to a factory and see a belt weighing system whose principle is shown diagrammatically in Fig. 12. The material is poured into a hopper and is dragged out of it at the bottom by a moving belt, and the thickness of the material is supposed to be controlled by a vertical shutter. The movement of this shutter is dependent on the weight on a pan beneath the belt. Subsequently the material falls off the thin inter-mediate conveyor belt on to the final one. This weighing system cannot be attached direct to the final conveyor as this uses a rubber band too heavy and stiff to allow accurate weighing.

The machine is started up and it is immediately apparent to you that the material is not only granular but slightly fibrous and sticky as well, and so the metering shutter does not produce an even top surface to it. The granules come out of the hopper in jerks as they are ploughed and torn about and sometimes half rotated by the edge of the shutter. The depth of the material is intended to be some four inches, but it varies constantly from $3\frac{1}{2}$ in to $4\frac{1}{2}$ in. Its top surface looks like a rough sea.

You gaze at it in horror. You have been told that a few minor adjustments are all that should be necessary and you are faced by a double intrinsic impossibility. The first is that the weighing system is weighing the average weight of material over a distance which includes several individual waves and is therefore quite unaware of them. The second is that even if the weigher spotted them it would be too late to do anything about it. The weigher is always reading yesterday's newspaper—busily correcting for densities that are already out of date.

You are taken out to lunch and everyone is anxious to hear you tell them how a 'few adjustments' or possibly 'some electronics' will solve all the difficulties. You have to break it to them that everything is wrong in principle. The weighing mechanism would work better if it did not work at all, for it is only throwing another imponderable into the system. You must not try to argue with intrinsic impossibilities. Everyone looks dispirited and, after a pause, you are asked what you would do. You reply that the problem can be solved but you must be allowed to start again from scratch, for you realize that there is nothing intrinsically impossible about the process problem as such, it is only the existing way of solving it that is hopeless. That someone has thought of the wrong answer does not mean that there is not a right one. They agree that you can go ahead. But how do you begin? To gaze at the problem as a whole is a most intimidating exercise. Well, don't panic, divide it up into bits and concentrate on each in turn. The obvious place to start is with the two impossibilities; first you must get rid of them to clear the air.

They are both caused by the fact that the material and the

machine are having a running fight. You decide to tackle the wavy surface first as this is the first bad thing that happens in the existing set-up. To do this you must make friends with the material; tune the design to harmonize with it. Therefore you must study how it behaves. Take samples and play about with them. Sit up on the machine and watch how the material alternately flows and clogs. Store your mind with accurate pictures; understand the characteristics of these awkward granules. Measure the various bulk densities yourself, it will help you to know the material better.

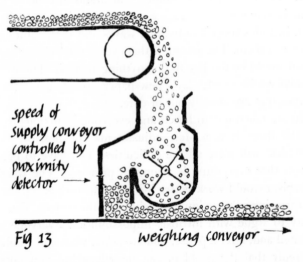

speed of
supply conveyor
controlled by
proximity
detector ⟶

Fig 13   weighing conveyor ⟶

Why does the vertical shutter make such a mess of the top surface of the material? There are two reasons, you find. The first is that the material in the hopper is lumpy in any case; then there is the ploughing effect of the shutter itself. And the second is made worse by the first. You can do nothing about the particle density but you can try to improve the bulk density. You experiment and find that if the material is dropped from a constant height through air on to itself, its bulk density will be much more uniform, but you must not drop it in lumps. Well, get the lumps out. You take a handful of it and squeeze it into a ball; it goes down to less than half its original volume, a soggy doughnut. You toss it up and down in the air, hitting it with your hand, and it gradually breaks up into granules. 'This is what we want,' you say to yourself, 'let us design something to do just that.' Put in a paddle that keeps lifting the granules up and dropping them. Then they must be pushed over a weir into a discharging section and fall through a constant distance to the conveyor band.

The volume of the material being supplied to this system must be controlled so that the dropping distance and the depth on the conveyor are constant. Put in a proximity device at the critical point and arrange for it to meter the volume of granules being supplied. It won't be perfect but you will have removed your lumps and taken some of the extremes out of the bulk density (Fig. 13). Now forget all about this problem and concentrate on the next one.

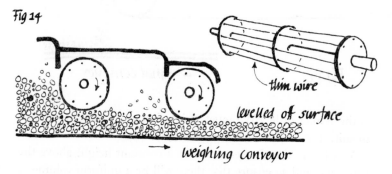

Fig 14

thin wire
levelled of surface
weighing conveyor

This is that the vertical shutter must be replaced with something more sympathetic so as to give a smooth surface and not a wavy one. You must not use anything that builds up a block behind it or exerts a vertical force; it must be something that flicks off or cuts off the surface material. Would a rotating wire cheese-cutter do? How about lengths of piano wire stretched tightly between thin discs mounted on a shaft and rotating very fast? You experiment and find that it is best to use two, the second of which does the final trimming (Fig. 14). You play about with the speeds until you find a combination that gives you a smooth surface. You can now vary the level of material off to any desired depth by moving these cheese-cutters up or down. This has therefore dealt with the first intrinsic impossibility. Now concentrate on the second one.

The basic problem here is that the weigher is always behind with weighing. To form a material first and weigh it afterwards

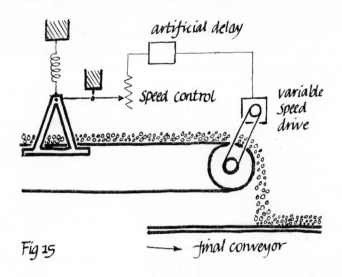

Fig 15   → final conveyor

is the wrong order. We must weigh first and form the volume to suit. This can be done quite simply.

You locate the cheese-cutters at a constant height above the conveyor, and so ensure that there will be a uniform volume of material per foot of the conveyor. Its weight per foot will vary solely with its density. You can now weigh it precisely as there are no waves. It follows that the weight on the pan is a direct measurement of the density. Next arrange matters so that the velocity of the conveyor belt at any moment is inversely proportional to the weight on the pan (Fig. 15). This means that the velocity of the belt will cancel out the density differences. And this is what we want. Or almost what we want, for it is not quite accurate. The granules falling off the end of the belt may not have exactly the same density as those being weighed at the same moment 3 ft earlier. You are reading tomorrow's newspaper instead of yesterday's. Well, you can correct that by arranging that your conveyor speed control has an electric delay system built into it so the speed of the conveyor is always behind what the weigher demands by the same interval that it takes the material to travel from the centre of the pan to the end of the conveyor. In this way the velocity of the belt at any moment would be tuned to the density of the material falling off the end of it at that moment. Good, for that's what we want; now for the next item.

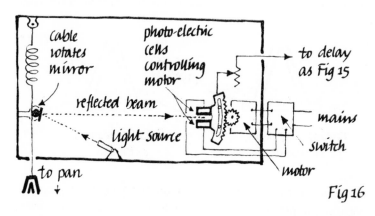

Fig 16

So far we have been making do with the existing weighing system. It should be highly sensitive and, as at present designed, it is not. We can reduce the friction by using better knife edges, nothing new in that, but the rest is more complicated. The position of the pan must operate a mechanism for adjusting the speed of the conveyor motor over a range of anything up to three to one and do this without any damping or backloading on to itself. Not only must friction be reduced to an absolute minimum, but inertia too. A vertical movement of the pan in the nature of one thousandth of an inch must be magnified sufficiently for it to alter the conveyor speed. If you build up the movement by a system of levers there is a danger that lost motion, inertia and vibration dangers will make the system too insensitive or jerky. What is the lightest thing you can use? Nothing can be lighter than a ray of light. You had better use one. A minute movement of the weighing pan can be magnified enormously by a mirror and a ray of light. Two photo-electric cells mounted one above the other will complete a servo system for controlling the conveyor motor via a delaying mechanism (Fig. 16). Forget it and tackle the next problem.

This concerns the indirect effect of the tension in the rubber conveyor belt itself. You have reached a degree of sensitivity in your weighing system where this tension will have a disturbing effect. Although your total pan movement is only a fraction of an inch even this small figure allows the tension in the band

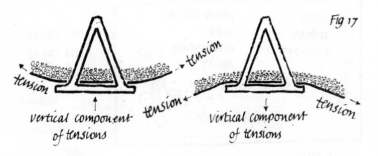

Fig 17

to influence the apparent weight on the pan, for it exerts a vertical component according to whether it is above or below the line joining the top surfaces of the conveyor rolls (Fig. 17).

If the tension in the band is always the same, you could compensate for this effect in your system, but unfortunately you cannot depend upon this. The longer the belt is used the more it stretches, and it must be re-tensioned periodically; in addition, the granules may tend to go where they are not wanted and jam themselves between the side of the hopper and the rubber belt and so introduce a further unpredictable load. Moreover, the power required to drag the material out of the hopper will not necessarily be a constant. You cannot avoid and dare not ignore any of these factors. You can think of only one solution. The pan must not be allowed to stray for long from its truly vertical mid position. This can be achieved by having a separate mechanism that lifts the whole weighing system up and down so that whenever the pan moves vertically it is brought

46

back again to the mid position. This can be done by arranging two sensing devices that operate an electric motor in the required direction (Fig. 18). Will this make the whole system hunt? No; so that's all right (see chapter 6).

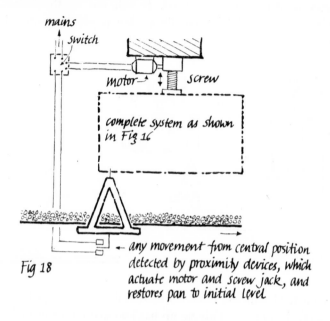

Fig 18

any movement from central position detected by proximity devices, which actuate motor and screw jack, and restores pan to initial level

Now you can sit back and have a look at all you have done so far. You have divided a major problem up into minor ones and solved each in turn. Now put all the solutions together (Fig. 19). Well, it should work all right, but it is terribly untidy and therefore expensive. Something must be done to tidy it up. You notice that the motor that moves the pan back into its mid position operates at precisely the same time and in the same direction as the servo motor for the weigher. You can therefore make one motor do both jobs and eliminate the sensing devices too.

Next, you must have a good look at that delaying device; it is expensive. Can't you find an alternative? You think about it and realize that there will be a small delaying action inevitably built into the system, for inertia will prevent the response of

47

the conveyor motor being simultaneous with the movement of the pan, and there will be other minor factors giving the same effect, such as the lag in the optical servo. All you need do is increase these slightly until their combined effect gives the degree of delay that you need. It will not be theoretically perfectly accurate for all conditions but it will be quite near enough.

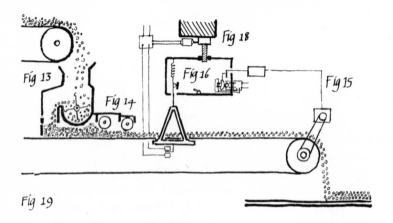

Fig 13   Fig 14   Fig 16   Fig 18   Fig 15

Fig 19

Your final solution, obtained by dividing up and tidying up, is shown in Fig. 20. Tidying up has knocked off a third of the capital cost and probably more still of the maintenance. Sometimes you have to go the long way round to discover the short way home.

Now we come to the fourth inventive principle, and this is of quite a different nature. So far we have been considering how we can design a machine to fit a given need. Now we will consider the reverse and try to find a need that will fit a given machine. I will call this principle 'feedback' and it can be illustrated quite simply. Some years ago a man invented a domestic washing machine for cleaning clothes. He built a prototype which was most successful in removing dirt but had one disadvantage. It tore the clothes into small pieces the size of threepenny bits. Those concerned with smashing up rags and paper for a board mill would love it, he thought. And they did.

48

It is a standard machine for this work and used all over the world. Its dimensions are scaled up many times but in principle it is identical to the original washing machine.

The magnetic and gyroscopic compasses were originally feedback inventions, and so are most forms of focusing electrical energy for metal removing.

This brings us to what is probably the most exciting and rewarding sphere of inventive design. It is the possibility of feeding back into practical exploitation, through engineering,

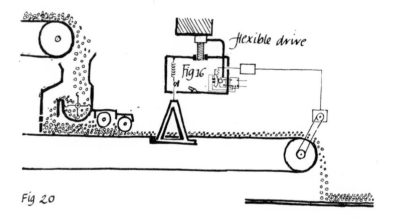

Fig 20

new discoveries in the physical sciences. One of the chief difficulties of doing this is that the engineer has often no regular contacts with scientists. He may feel that it is more than likely that some recent scientific advance could well help him in solving a particular difficulty, but he does not know how to find out what or where it is. On the other side the scientist may also be convinced that some physical phenomenon must have a useful application but he, too, does not know where. Many of the larger firms want to speed up this intercommunication between designers and scientists by creating special departments for doing it. Personally, I think it would be better if some central register of new scientific phenomena could be established where computerized cross-references could be available to both parties.

This chapter cannot end without saying that, useful as these principles may be as guides to designing, it must never be forgotten that the inventive element is essentially independent and unforced. Subsequently disciplined or not by art or logic, it initially stands in its own ground and owes nothing to its neighbours. Expert experience in the field is not essential; in fact, too great a knowledge of existing techniques may be a handicap. The following examples of inventions and inventors could be greatly lengthened, but should be sufficient to illustrate that a formal education in the field of the invention is not always essential.

Invention	Inventor
Safety razor	Traveller in corks
Kodachrome films	Musician
Ballpoint pen	Sculptor
Automatic telephone	Undertaker
Parking meter	Journalist
Pneumatic tyre	Veterinary surgeon
Long-playing record	Television engineer

The inventive is the most fascinating and most fickle of a designer's muses; she sometimes ignores professional qualifications and mocks the conventionally minded.

These guides to inventiveness, which are signposts and not moving staircases, can be summarized as follows:

(1) Concentration and relaxation
(2) Do not be conditioned by tradition
(3) Complicate to simplify
(4) Make allies
(5) Divide up and tidy up
(6) Feed back from physical sciences
(7) Don't despise the untutored inspiration

4

THE DESIGN OF DESIGN
THE ARTISTIC

Some people are tone-deaf, others are colour-blind, and there is little to be done to help them. But if a person has an ear for music or an eye for colour he can have his faculties disciplined and trained and as a result find them more enjoyable and useful. Similarly, a sense of engineering style, once there, can be developed and refined. If it is not there at all there will probably be a compensating extra sensitivity in other fields, as is often found in the tone-deaf or colour-blind people.

If the introductory examples of style which were given in chapter 2 were entirely incomprehensible, do not worry. Possibly my explanation of it was inadequate, so we will try again in more detail.

As already explained, defining style in words tends to be unsatisfactory in most realms of art, literature and music, and this applies particularly to engineering, and so the problem of presenting it in the form of general principles is a rather formidable one. I will do my best.

Perhaps an easily recognizable artistry of engineering lies in the continuity of energy. The contrast between the piston engine and the jet engine for aircraft propulsion is an example of it. We all know the technical reasons for adopting the latter but fundamentally it is a matter of style. I well remember when this fact made its first impact on me.

I was lying on my back on a Surrey hillside on a hot Saturday evening during the last war and watching the flying bombs passing overhead. Not only were there flying bombs but fighter aircraft were running after them trying to shoot them down. The infinitely more powerful and expensive fighter was only fractionally faster than a flying bomb. The difference in engineering style was painted on the blue background of the sky. The basic distinction between a piston engine and a pulse

4-2

jet concerns the utilization of energy. The little flying bomb portrayed continuity, while the fighter was always interrupting the flow of energy. Every engineer is aware of the basic principle of the internal combustion engine but its style may be less apparent.

The energy source lies in the expanding gases which push a piston down a cylinder. But what does the piston push on to? The usual reply is the connecting rod, but this is not so. The piston pushes on to an oil film. Then the other side of this oil film exerts a force on to the connecting rod. The energy has to

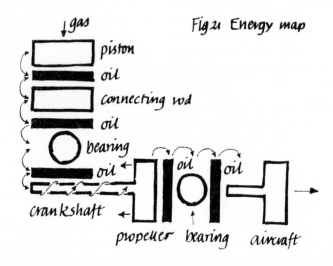

Fig 21 Energy map

go from gas to metal, then to oil, and then to metal again. Its continuity has been interrupted three times already. If we assume roller bearing big-ends, the energy will have to cross another three frontiers before reaching the crankshaft, i.e. oil–metal–oil–metal. The crankshaft is attached to a propeller which dissipates all the energy into the air. The reaction from the air pulls the crankshaft along. This pull is transmitted to the crankcase by radial and thrust bearings. You then attach an aeroplane to the crankcase. The overall energy route is shown in Fig. 21, and is one of gas–metal–oil–metal–oil–metal–oil–metal–oil–metal–oil–metal.

Not only has the energy had to negotiate this hurdle race but having reached the other end some of it has to come hurdling back again, for any one piston is absorbing energy and not giving it out for three-quarters of the time. In addition, the direction of the energy is changing from the vertical to the angular, and the angular to the horizontal again. It is a hotbed

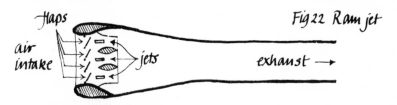

Fig 22 Ram jet

Sequence of operations
1. Air flows in through flaps and mixes with fuel from jets
2. Mixture exploded by heat from walls, flaps close
3. Exhaust out of open end
4. Due to 'organ pipe' resonance, wave of reduced pressure travels back into nose and sucks flaps open again.

energy map

A 'V1' jet gave twice the effective HP of a modern 3-litre F1 Grand Prix racing engine, and weighed less, and used inferior fuel.

of discontinuous energy and no wonder it is wasteful. Compare all this to the flying bomb with its pulse-jet principle (Fig. 22).

Here all the energy is horizontal. There are no oil bearings to hop over. The gases push the bomb along at first hand. Simplicity and continuity is its style and a good one too. This principle always applies to designs where energy is originated or absorbed, even though it is not a machine as such. Any load-bearing structure must deflect to some extent and therefore must absorb energy. High-tensile bolts uniting plates by friction are better than normal bolts or rivets, and good welds are best of all.

The next suggestion with regard to style is one which, although negative, is also of wide application. You must not over-design. Many a good idea has failed to keep its initial promise because the inventor was so enthusiastic that he exaggerated it. It became his master instead of his servant.

One of the worst examples of over-designing I have ever seen I did myself. My only excuse was that I was a first-year under-graduate at the time and knew no better.

I was very interested in the problem of how to make a racing car go round corners quicker. I came to the conclusion that the control of skidding was a major factor. To control a car at speed it ought to be possible to make either the front or the rear wheels skid sideways at will and, if necessary, both at once. It was not difficult to design a car that would do this. If the drive was to the rear wheels and the suspension, weight distribution and tyres were suitably arranged, then the front wheels would skid rather more easily than the rear if the car was coasting. When the power was full on, the rear adhesion would be partially diverted to transmit it and so the rear wheels would slide side-ways before the front ones lost their grip. This was the orthodox technique, but it had one major difficulty. With the rear wheels skidding sideways the car would fly into a spin unless the skid was immediately and accurately corrected by turning the front wheels so that the front and rear of the car maintained the same radial velocity outwards. This is difficult to do, especially at high speed and on bumpy roads when the car may jump side-ways at the rear far more quickly than the front wheels can be steered to match.

At that time the solution appeared to me to lie in reversing the driving characteristics. By transferring the drive to the front wheels they could be made to slip sideways when required by turning on the power. The rear end could then be designed to slide first under all other conditions by arranging the type of contact area, distortion angle, and weight distribution so as to produce this effect.

One would then have a car that would have to be driven differently, for power on would mean skidding at the front and

54

not at the back. But this would bring with it a useful characteristic, I thought. The car would be essentially more stable under power, for the skidding and correcting would both take place at the same point of contact with the ground, i.e. at the front wheels. It would be inherently stable and self-correcting. One would merely have to turn the front wheels to a greater angle and this would be done automatically. There was one obvious difficulty. Weight transfer on acceleration would minimize the grip of the front wheels and power would be lost through wheelspin. As much weight as possible must therefore be built over the line of the front wheels. The reasoning up to this point was quite sound and is the basis of the remarkable road-holding and racing performance of the Mini-Coopers, which appeared over a quarter-century later.

Unfortunately, I then started to over-design with a vengeance and exaggerated the idea out of all recognition. To ensure that the rear wheels skidded first when the car was coasting or braking, I increased the area of contact at the front by fitting two tyres side by side on each front wheel. This was wild enough but there was worse to follow. The centre of gravity was to be as far forward as possible, so I was determined to sacrifice almost anything to get this. Every movable accessory, such as the petrol tank, was put in front of the front axle. Then I turned my attention to reducing the weight of the rear axle. I came to the conclusion that a substantial part of the weight of an independent rear springing system was due to having to withstand the braking torque. Well, I would dispense with the rear brakes. This would eliminate the fore and aft forces as well as the twisting. All that would now be required of the suspension was to support the weight and the sideways friction forces. This simplified the suspension design a great deal but then I decided to go still further. I left it off. I made the rear chassis members much thinner than they should be so that they could act as springs, and fastened the back wheels directly to them (Fig. 23). Everything was in turn over-designed.

I built this exaggerated machine and entered it for a two-mile hill climb near Belfast. As I only finished it a few minutes before it had to leave to catch the boat, I never had an opportunity to test out the handling in advance.

A few days later the car was unloaded in Belfast and towed to the paddock to be inspected by the stewards of the R.A.C. I immediately asked them to gather round while I demonstrated the front brakes which were operated through the transmission. If you clamp on a transmission brake hard when there is an unevenly wet surface, it often happens that one wheel spins backwards due to the action of the differential. This greatly

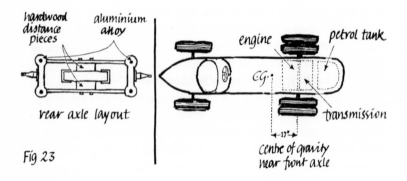

hardwood distance pieces   aluminium alloy

rear axle layout

engine   petrol tank

CG·

transmission

centre of gravity near front axle

17″

Fig 23

impressed the officials, who then departed, fortunately overlooking the fact that the car had no rear brakes at all. Now all I had to do was to drive it fast.

As soon as I left the starting line I discovered that over-designing is not a good idea, especially when you have to drive up a hill you have never seen before and are uncertain which way the next corner goes. After my first run a committee of all the officials was called and, after a solemn debate, I was informed that I would not be allowed to drive up the hill again. The secretary said, 'We are used to street fighting and things like that in Belfast, but we cannot face a repetition of your first performance. It is not that you came out of all the corners sideways; you went into them sideways too; in fact it is debatable whether your front or rear wheels first crossed the finishing line.' I pointed out that I was second fastest and thought I could do much better next time, but all to no avail.

The basic idea of built-in skid correction by combining the driving and steering wheels was a sound one and worked well as such, but any gain in cornering time was more than counterbalanced by the phenomenal over-steering on the overrun. The exaggerated design caused an exaggerated transition from over to under steer which was complicated still further by the hopping about of the rear wheels which made cornering at 80 m.p.h. or over rather uncertain on bumpy roads in the wet.

Nearly always in the evolution of a new invention there comes a stage when there is a strong temptation to over-design. The inventor is so enthusiastic about his idea that it numbs his critical faculty. It is here that a feeling for style is a real safeguard. And even the word 'over-designing' may alert your mind to be aware of this most tempting tendency to exploit a design so much that you spoil it.

A further criterion of style is the rational. Here again, definitions are difficult. It is more obvious than useful to say that a machine with a rational principle of working is better than a totally irrational one, for the latter would be unlikely to work in any case. It would be better to say that the more rational the neater the style will be.

As already pointed out, the design of the traditional cartwheel has a rugged and durable simplicity, but a bicycle wheel has much the better style. It rationally combines and exploits two facts.

The spokes of a cartwheel have to withstand cornering forces as bending stresses and the vertical load as compressive and bending ones. They have therefore to be of substantial cross-section.

The bicycle wheel absorbs all cornering and loading forces as pure tension in its spokes and, in addition, transmits a driving torque from the hub by mounting them tangentially to it. Linked to this is an appreciation of the fact that steel in tension can be made of hard-drawn wire, thereby increasing its strength several times for a given weight. It is a neat partnership of materials and mathematics.

In principle the same contrast of style between the bicycle

and the cartwheel can be seen side by side in the railway and road bridges over the Forth. The thickset tubes of the former contrast with the wire-suspended economy of the latter. The Severn Road Bridge has developed this style even further (Fig. 24).

I found a simple example of the value of the rational in a process mixing problem. A quantity of ingredients had to be

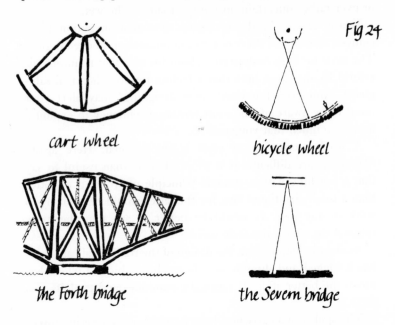

Fig 24

cart wheel

bicycle wheel

the Forth bridge

the Severn bridge

mixed together as thoroughly as possible, but it was important that the temperature should not exceed a certain limit. All the conditions could be tightly controlled except the consistency of one of the ingredients, where the viscosity varied a good deal. If a fixed time for mixing was used the more viscous examples drove the temperature up too quickly, and the less viscous were undermixed. The operators were testing the temperature of each batch as it came out and adjusting the mixing time accordingly, optimistically hoping that the next batch would have similar characteristics to the one before it. It often did not, and so there was no real rational basis. A simple solution was to recognize that the temperature increased solely as the result of the work done in the mixing process. A wattmeter to cut off the driving motor after an appropriate degree of energy had been dissipated was all that was needed.

Sometimes the solution is not quite so self-evident.

A firm had to dry many tons of granular material per hour and the main problem was the unpredictable amount of water in it. As the material was bought by weight the suppliers were not very worried if a lot of water was present and I sometimes wondered if the sacks didn't spend the night in a local pond before being delivered. The water content varied from 12 to 45 per cent and this figure had to be reduced to a consistent 10–10½ per cent before processing.

The granules were tumbled through a horizontal revolving vat dryer where hot air was blown through by a powerful fan. The heating was produced by an oil firing system and could be varied instantaneously over a wide range. The difficulty was that no agreement could be reached on how much heating was needed. One man was trying to judge by hand the water content of the granules going in and turning up the oil burners to suit. Another man was taking samples from the output end and usually wanted some other setting. Generally both views were wrong. There was no rational basis to the system. The problem was to find an automatic and accurate control for the temperature of the air entering the dryer so that at any particular moment the granules were just sufficiently dried so that they would leave with 10 per cent of water, quite independently of how much water they had when they went in.

Any rational scheme would have to depend on some factor that was both relevant and measurable. In the whole system only one thing fulfilled these conditions, i.e. the temperature of the air leaving the dryer. If there was a constant volume of hot air entering the dryer this final temperature was reduced by the amount of water that was evaporated en route. The greater the temperature loss the wetter must be the material in the dryer, and so the higher the input air temperature should be. This could be arranged quite simply, by regulating the oil burner automatically so that the temperature of the air leaving the dryer was always the same. If it showed any signs of dropping the thermometer would give a signal that would increase the temperature of the air entering the dryer until the required exit temperature was reinstated. In this way the input air temperature would increase linearly with the water content of the ingredients, which was exactly what was needed (Fig. 25). Once the output temperature had been linked to the water content by experiment a suitable setting could be fixed and the machine could be left alone. The rational had replaced the irrational.

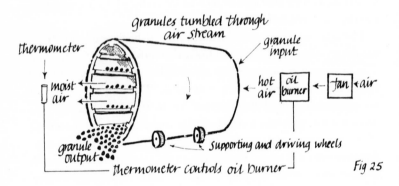

Fig 25

The fourth criterion of style is that of choosing the best medium in which to exploit an invention.

An engineering device which is now standard practice and universally used for certain purposes utilizes a principle which is twice as old as itself. This is that of using a variable angle swash-plate to give an infinitely variable speed drive. Originally, for over a quarter of a century, designers toiled patiently and unsuccessfully to make a practical proposition out of it. All they achieved was excessive weight and unreliability. Their mistake was that they tried to exploit the principle mechanically instead of hydraulically. Today hydraulic drives are an automatic choice in many instances. They are theoretically ideal for low speed and high torque transmissions, such as those needed in earthmoving vehicles, as well as a host of industrial applications.

Choosing the right medium revolutionized everything. Style is vital. To portray a rainbow you should paint it in water colours rather than model it in clay.

So important is this question of style that it is sometimes best to decide the medium first and do the inventing afterwards. Indeed the machine may almost invent itself once the style is right.

For instance, if you had to design a device that would lay or pick up road-marker flags at speed you would first think about

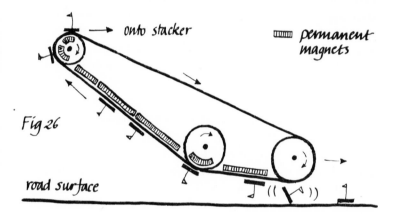

onto stacker   ⬜ permanent magnets

Fig 26

road surface   (( ))

the medium. Grab them mechanically? Suck them up? Use some kind of hook? None of them seems to be very promising. How about permanent magnets? This starts you off thinking and quite quickly you arrive at the answer (Fig. 26).

Now we must turn to the difficult topic of assessing the degree of permanence in style.

As already explained, the appreciation of style in many art forms may be more instinctive than logical, but in engineering the practical importance of the right verdict is greater than in any other realm. And, in one sense, more complicated too, although at first sight it might well seem simpler because of its more direct relationship with the imagination.

The imagination of a poet paints an intriguing or emotional picture, and he then must search for words to interpret it poetically.

The painter is better off, or he can reproduce an imagined or actual scene, or a combination of them, by portraying in two dimensions what he sees, or imagines, in three. The engineer both pictures and creates in three dimensions. Unlike his fellow artists, he need not interpret, his dreams literally come true.

But, also unlike his fellow artists, he has the added complication of assessing whether a style is transitory or fundamental. That a style is good does not mean that there will be none better. He must decide whether to polish and refine an existing design or look for something entirely new. It is easy to be wise after the event and say, for instance, that money should have been diverted earlier for the development of the jet engine. The important thing is to be wise in advance and make events conform. But it is by no means easy to decide whether a style is potentially obsolete. If there is a general answer to this problem I do not know it. But I think I know one clue to what it might be.

If the design of a particular machine or production line is based on the way the process was originally done by hand, it is unlikely to be the final form. The feeding forward by mechanizing a style that was handy when only hands were available is doomed, in the long run, to be superseded by the feeding back of ideas and materials from the physical sciences. Beware of well-dressed arts and crafts.

For instance, since before the beginning of this century, the style of paper-making machines has remained basically the same, altering only in scale and speed as it is clothed with increasing technical sophistication. But underneath it all there lies the old hand-made paper principle. Is it permanent? I doubt it. A style which mixes all the raw materials with twenty times their own weight of water at great expense, and then removes it again at even greater expense, is unlikely to remain economically ideal, even although the water has a molecular binding effect.

On the other hand the mechanized coal mine has wisely forgotten that picks and shovels ever existed. The traditional design of aircraft has a style nearer to the hand-thrown boomerang than to the internally powered swing-wing seagull. Perhaps, after all, Mr Roe could have learnt even more on his seaside holidays. Again, surely the resonant pile-driver, with its principle fed back from applied mathematics and physics, must ultimately kill the traditional 'hammer and nail' style of the old techniques, and there will be no mourners at the funeral.

This chapter might be summarized by saying that a sense of the artistic in engineering style, as in other fields, may be confused rather than clarified by attempting definitions, but, despite this risk, the following principles are worth remembering:

(1) Aim at continuity of energy
(2) Avoid over-designing
(3) Choose a rational basis
(4) Find the appropriate medium
(5) Avoid perpetuating arts and crafts

THE DESIGN OF DESIGN
THE RATIONAL

The rational element in creative design is tending to divide into two distinct spheres of thought. For convenience of discussion we might call them 'computer logic' and 'human judgment'.

The use of computers as tools of design has opened a new horizon of possibilities, which we have hardly begun to exploit.

First there is the direct economic benefit. A few years ago it was thought inevitable that a draughtsman's time would be much occupied in routine or involved calculations. Now, or soon, a computer will do them for him. And it can do more. It can design economically in three-dimensional problems where the sheer volume of calculations needed makes their human assessment impossibly protracted. It can be logical where we have only time to guess.

But there are many indirect benefits too. It is a common experience that when a newly designed machine first starts up, trouble, if it occurs, is found to be due to one of two reasons.

The first possibility is that the underlying principle of the machine may be wrong. The second, and much more usual difficulty, is that the detail design of some minor component is faulty. How often has a designer seen, with horror, as fitters are assembling his new machine, that some parts are clumsily or superficially designed. To redesign and replace them would delay the starting up of the whole project, but to leave them as they are might waste even more time and money in the long run.

These highly irritating mistakes in detail designing are sometimes the outcome of economic pressures. Many machines have a multiplicity of small components and the cost of drawing them down may be a significant item in the total expenditure. As there is nothing inherently difficult about designing them, the lowest-costed, and therefore the youngest, draughtsmen are given the job, leaving their seniors to concentrate on the major and self-evidently critical assemblies.

But experience in detail design, which is precisely what a young draughtsman does not have, is something which small components need as much as big ones. A computer can often be taught quicker than a draughtsman and does not forget things so easily.

But there is a further disadvantage in non-computerized routine designing.

A young man may be at his best age for innovation and invention, and find it unbearably frustrating to be condemned to endless detailing. Of course, a knowledge of machine drawing, and the discipline of doing it, are an essential part of any engineer's training, but to give him little else to do may drain away his inventive enthusiasm.

One sometimes finds people who delight in the meticulous designing of details, but they are as rare as they are valuable.

But there is still another way in which the commonplace may be worse than tedious. It may be design-destroying as well as soul-destroying. The new idea, embodied in a compatible style and clearly sketched down by the inventor, may start off well but then, gradually, the original clarity of thought may be lost in a blur of detail. A computer will do what it is told. A draughtsman can easily be tempted into making minor alterations because it may be easier to calculate or draw that way.

Computerize all that can be conveniently done and release young minds more for the adventure of originality. You can keep the wood in sight if you don't get lost in the trees.

But there is much that a computer will never be able to do. Only human judgment can take the responsibility for many logical decisions. A computer has no independent logic of its own. It is a parrot, not an owl, and by teaching it we teach ourselves too; programming is a good mental discipline, as it soon reveals fuzzy thinking.

In this chapter we will concentrate solely on those principles of design that are inescapably the personal responsibility of the designer himself, and probably always will be.

The first of these principles is to settle down and work at the problem. The vast majority of designing is a matter of logically thinking things out. As has already been mentioned, logic is the element that the designer shares most closely with the machine. Problems that at first sight appear to call for some new invention to solve them are often found to disappear when dissected and considered with care. All the parts of many machines and some parts of all machines are best designed by simply getting down to designing them.

Take, for instance, the patterned floor covering machine we have already twice used as an example. Once the general principle had been laid down it was a matter of logic to fill in the details, even though some sections had the appearance of being highly demanding.

It will be remembered that one of the specified features of this machine was that it should run continuously. This meant that when one pre-laid stack of assembled pieces, about 6 ft × 3 ft, had been reduced to one or two layers another stack had to be substituted for it. This could only be done during the interval the suction box was away on the laying section of its cycle. When it came back the new stack must be there waiting for it. The exchange had to be completed within 4·8 seconds.

During this interval the platen which carried the now exhausted stack had to be removed from the machine sideways over a distance of 10 ft, a dimension fixed by the necessity to have sufficient clearance for a platen-changing system. Next the new stack on its platen had to move sideways 10 ft and be located in every direction to $\pm$ 0·002 in. This included the height of the top surface, which had to be the same as that of the last used layer in the exhausted stack. If, for some reason, the outgoing stack had to be removed before it was fully exhausted, or possibly hardly used at all, the new stack had still to take its place accurately.

In addition, everything must be done without jerking, as this would disturb the pieces composing the stack. The platen with its load and supporting mechanism would weigh about half a ton. Moreover, any mechanism for the vertical movement or

66

location of the stack must be so constituted that it would also lift the stack by a distance equal to the thickness of one layer at each cycle during its running life. This incremental height must be controlled by the actual thickness of the particular layer, not just the nominal one, as otherwise an accumulated error might build up. Of course, all this mechanism must also be as light and simple as possible and remotely controlled, as it has to move in and out with the stack (Fig. 27).

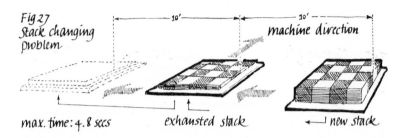

Fig 27
Stack changing
problem

machine direction

10'

10'

max. time: 4.8 secs   exhausted stack   new stack

On the face of it, this total stack-changing problem seems quite difficult, but it is not when approached logically. You do not need to invent anything, just think it out. It is all quite easy.

Begin by having a preliminary look at one dimension at a time. First we must move the old stack out 10 ft sideways and then the new one in to take its place. But why do it in two stages? Why not make a combined transporting unit that will move one out as the other comes in? That's a good idea. We'll make one 10 ft moving mechanism do both jobs, at the same time.

Now we will look at the vertical dimensions. How about fixing a vertical dimension and making all horizontal surfaces conform to it?

If we fixed this at, say, 3 ft 6 in. from the ground we could make everything level with it. The suction suckers would always reach down exactly to it during the pick-up period. The top surface of a new stack would already be lined up at this height before going into the machine so that no further vertical movement would be necessary during the critical 4·8 second period.

67

As the stack has successive layers removed arrange a team of photo-electric cells to look through slits and move the stack up each time so that top surface is always at the fixed height.

Fix co-ordinates for the other two dimensions and see that the new stack lines up with them when still outside the machine. Move it in accurately over the 10 ft and that's all you need do. It will not require any other movement to ensure its specified three-dimensional location with the departing stack.

Run the whole thing in on precision rails, with wheels in two planes, until it ends up against non-resilient buffers and then

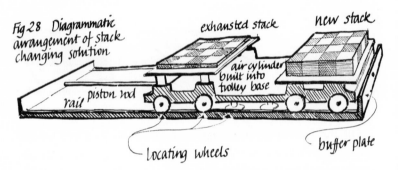

Fig 28 Diagrammatic arrangement of stack changing solution

exhausted stack

new stack

air cylinder built into trolley base

rail piston rod

locating wheels

buffer plate

hold it there. We now have automatically the accuracy we need and the only chance of going wrong is in human process of building the stack to the right height. This can be safeguarded by having 'Go' and 'No go' gauges (Fig. 28).

Having fixed all our accuracy in advance, so to speak, we can now tackle the design of the movements. First, we have to move over half a ton sideways 10 ft without jerks and stop it without impact in 4·8 seconds. What shall we select to do it for us? Which medium—mechanical, hydraulic, electric, compressed air, or a combination of them? As we must design for minimum shock we want minimum acceleration. All these mediums can give this to us but the cheapest will be air. A constant pressure will give a constant force on a piston and constant acceleration to a mass, and the same characteristic on braking.

Apply this principle to moving the trolley changing the two stacks by using a 10 ft stroke cylinder and piston so arranged that the first 5 ft are spent accelerating and the next 5 ft braking. Put 80 lb/sq. in. on to the piston through relatively large pipes and ports and the acceleration will be largely uniform. Brake by putting 160 lb/sq. in. on the other side of the piston. When it has almost stopped, release the 160 lb/sq. in. side so that the 80 lb/sq. in. will push and hold it firmly against the buffers.

There are many other ways of achieving the same result. As I have already said, you don't have to invent anything new.

To complete the design we must arrange for the vertical incremental movement that is controlled by the photo-electric cells looking across the surface of the stack. There are in-numerable ways of jacking things up; we want the best combination of lightness, cheapness and precision. Four air cylinders, one at each corner of the 800 lb stack and platen, can provide the force and be easily fed by a flexible hose, but by themselves they are hopeless at keeping in accurate step with each other and we must keep the stack level.

What is the best medium for precision of movement? How about electricity? Quite a large motor would be necessary and its inertia would be a nuisance and also the total weight with gears and screw jacks would be high. Air seems better. Suppose we split the problem up; use one medium for the lifting force and another for the incremental control. Now we are getting somewhere. Use the air cylinders for the vertical force and let their movements be limited by a fractional horsepower electric motor. One motor would control all four cylinders through ganged racks and pinions. The air pressure would both push up the platen and act against the irreversible multi-reduction gearing of a little electric motor, which in theory would act as a dynamo but in practice have no load on it at all, and so be a good metering device. Thus by mixing our mediums we achieve the best of both worlds.

We have now attained our overall purpose and it has been done in unspectacular and logical steps. When faced with a problem this procedure is the one you should try first. Most problems can be solved by keeping your head, and using it.

The next principle in the realm of the logical is that money is

69

valuable. Most designers would like to feel that money was something that could be ignored except in so far as their salaries were concerned. Many creative artists are temperamentally vague about finance. They don't mind living in garrets. Unfortunately shareholders object strongly to living in them and shareholders have to be kept happy. They supply the money and they want dividends. No designer can escape from the discipline of the economic, and no one else in business can either, or not for long. Managements sometimes want to forget this too. I remember the nasty looks I received at a board meeting by quite innocently remarking that a factory building should be a mackintosh for a machine and not a monument to the management.

Now one of the cheapest ways of designing something is not to design it at all. Use one that is designed already by someone else.

I once saw a carpet-manufacturing machine where a very large number of connecting rods were needed to convert rotating motions into reciprocating ones. When the machine was operating they went so fast that all you could see was blur. When it stopped you could see that they were Ford Zephyr connecting rods. Readily available, precision made, no designing time needed, and above all, mass produced and so good value for money.

A draughtsman was once struggling with a machine-actuating layout which needed a fairly involved hydraulic set-up. I happened to spot that the hydraulic principle was identical with one used on a sophisticated fork-lift truck. Although the machine being designed had a quite different purpose and appearance, the same hydraulics would do, and they did. Once again designing was eliminated and proved components and locally available spares all came in as fringe benefits to a capital outlay reduced to a fraction. Shop around in other industries and you may find cut-price bargains.

If you have to design from scratch, remember that what is normally required is not the best possible design but the cheapest that will work. The temptation to design something that is new or sophisticated merely for the fun of doing it is a very real one and the more interested a man is in design the harder he may find it to resist.

Take, for example, the case of a long line of water-cooled tanks, each designed as in Fig. 29. The contents have to be maintained at a temperature constant to ± 5 °F, by means of controlling the input of cold water to the surrounding jackets. Chemical reaction of the contents produces a small and fluctuating amount of heat for about 10 hours and hardly any cooling

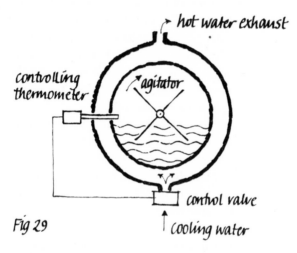

Fig 29

water is required. Then the heat output rises sharply until the whole of the cooling facilities are needed. We must install some kind of temperature-controlling system. Experience shows that spray cooling is not sufficient in the final stages as only a solid jacket of water can stop the chemical reaction running away with itself.

There is a perfectly standard solution to this. A thermocouple would give a signal which, when amplified, could be so adapted to control the temperature without appreciable hunting. With its water valve it would cost several hundred pounds per machine.

If we used a simple 'on and off' thermostat, where the water valve is operated by an expanding gas thermometer, we could do the job at a tenth the cost and less than a tenth the maintenance, but the temperature would hunt to an unacceptable $\pm 20\,°\text{F}$ and that is hopeless.

But are we sure that we are not dismissing a crude design like this too quickly? Possibly some technical snobbery may be unconsciously influencing us. Is it quite impossible to improve on this $\pm 20\,°\text{F}$ range of the direct-operating thermostat? Is it really hopeless? Have we tried hard enough?

It is not really very difficult. All we need to do is to use warm water instead of cold for all but the final stages. We can maintain a supply of warm water at any chosen temperature by feeding in some hot water from the exhaust main. The system will still hunt, but only within the $\pm 5\,°\text{F}$ that we are allowed. Plebeian it may be, but it is cheap and it works and these are the things that count in design. Pretend that it is your money that is being used; it is a good discipline.

This brings us to a point where we can ask a general question but cannot give a general answer. 'Is a cheap design with a predictable life better than an expensive design with an indefinite life?' It may well be. It often is. If you can be certain a component will last, say, 18 months you can arrange to replace it every 12 months and you never have a breakdown. Use a more expensive design that might last three years or ten and you will probably find it lets you down suddenly and disastrously. And its expense and long life have made it uneconomic for you to tie up capital in carrying spares, or anyone else to do it for you. Design a machine to wear out and it never will, for you can plan and afford the maintenance. Design it to last for ever and it certainly will not. Although there are exceptions, this is true in more instances than is often appreciated.

This question of maintenance brings up a point which is not always remembered. If a component is known to need periodic replacement you must be able to do it quickly. There is a certain make of car, a 1929 model, where you must first remove the engine before you can reline the brakes.

The necessity for rapidity of replacement becomes more and more vital as the day of the single machine disappears. The future lies in a series of machines, preferably spanning in one line the whole operation from raw material to packing. The problem is that this principle can nullify all its direct economic advantages in labour production costs if it is susceptible to long delays for maintenance. When one section stops it all has to stop.

I knew a factory where a new production line costing nearly a million pounds in all was at the mercy of one component in one section. This component was known to wear steadily in smaller machines. In the new line it wore much faster as speeds and pressures were correspondingly higher. But the machinery was much heavier too, and it took a gang of 10 men over a week to replace this one component, despite weekend working and night shifts. It made the whole new project uneconomic and the section concerned, which cost £250,000 to build, had to be re-designed, at great expense, so that replacements could be fitted quickly.

Not only do machines wear out but they can also smash themselves. I sometimes think that an electrical engineer has an easier life in this respect. Apparently everyone expects his machine to go wrong. He covers it with overload devices and no one minds paying for them, in fact they are legally necessary in many instances. Actually his type of energy is more deter-minable and confinable than that of the mechanical engineer, who cannot bypass his forces to earth in an emergency. His escaping energy often chooses its own direction.

I was once asked to safeguard the future working of a giant hydraulic press which had broken down. The basis of its hydraulic system was that the heavy downward pressure was provided by large-diameter cylinders but the quick-acting re-turn was operated by two smaller-diameter ones. One day both downward and upward circuits operated at the same time. Something had to give. It was the piston rods on the small cylinders which stretched and finally broke. The two pistons flew up their cylinders and hit the ends off them. This hardly slowed them down at all and they went straight through the factory roof, leaving two circular holes neatly punched out.

73

They then continued going up. No one ever saw them again. They may still be going up.

It is wishful thinking of the most dangerous type to assume that relays or switches, of whatever medium, are never going to fail. You must make sure that when they do, nothing too disastrous will happen. It is easy to talk about 'failing safe' but it is much harder to achieve.

All engineers should keep in mind the story of the small boy who went into a shop and asked for a package of detergent. 'What do you want it for?' asked the proprietor. 'To wash my budgerigar in', said the boy. 'That won't do it much good', replied the man.

Next day the boy came back. 'How is the budgerigar?' said the proprietor. 'Dead', said the boy. 'I warned you', said the man. 'It was not the detergent that did it,' replied the boy, 'it was the wringer.'

Designers, too, often fail to recognize where the real danger may lie. They may even increase it by over-protecting elsewhere.

After a rather hectic racing season in the front-wheel-drive car already described, I decided to use it as a road vehicle and enter it in the London–Exeter Motor Trial. For the rough surfaces of this event an adequate rear suspension system would have to be fitted and I installed a reversed quarter-elliptic layout (Fig. 30). As the steering lock might prove inadequate for sharp hairpin bends I decided to have a powerful and independent rear braking system, operated by a hand brake. This could be used to bring the tail round when required. I made the brake-actuating system as robust as possible; even the brake handle was a forging. I felt that if I needed the brakes at all I would need them badly, so nothing should be left to chance. The brake cables were oversize and the drums colossal. As the car had no orthodox electrical system and there was no time to fit one, I had to wire bicycle lamps to the mudguards for use at night. These soon objected to the vibration, went out or fell off. I continued without them. Coming up to a corner fast, at about 2 a.m., the moon suddenly went behind a cloud and

I pulled hard on the handbrake to set the car up in what I thought might be the right direction. The massive brakes bit hard and immediately the torque reaction member bent and broke at the point marked A. The unrestrained back axle instantly rotated and wound up the brake cables round the axle like a winch. Next the handbrake cut off its stop like a shearing cutter and went straight on to and then into the road. Amid a shower of sparks the car stood up on the end of the handbrake, which ploughed into the road until we left it and disappeared into a field: a silvery ghost tilling the soil of

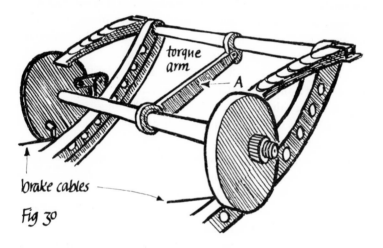

Fig 30

Somerset at unearthly speed. I had forgotten that too much strength can be a bad thing. It transfers energy rather than absorbs it. If you can, always build safety into the system rather than tack on warning or protecting devices later.

For instance, if you wish to trigger off in timed sequence a long series of operations A, B, C, D..., etc., you have two basic methods of approach. You can arrange matters so that when A is complete it actuates B and later B actuates C and so on, or else you operate each independently of the other by a camshaft or equivalent mechanism. By the first method everything is interlocked, in the second separation of control ensures that local failures have only local effects. Which is best depends entirely on the circumstances but it is a decision that must be made early in the designing.

Today you can buy, almost off the shelf, torque-limiting couplings, load cells and transducers of all types, which can be unobtrusively built in and arranged to shut off the power quicker than any human reaction. If, even for a moment, there is a chance of energy escaping and running riot, absorb it in distorting steel. Mild steel is marvellous stuff for doing this. Design so that the working stress is just below the yield and give it elbow room, and your errant energy will harmlessly dissipate itself in non-reversing plasticity. It is most important to have all these insurance factors, whether of the cut-off or escape-road type, built into the original design. You may not have the space to tack them on afterwards.

The total expense of insuring against escaping energy in these ways is often less than that of even one of the breakdowns which might otherwise be inevitable in the long run.

This brings us to another possibility which can occur in many process engineering plants.

If, for some reason, the whole production line, or part of it, has to be stopped, we must consider the fate of the material already partially processed within it. It depends, of course, on the physical characteristics of the material. If it will spoil after a certain time, cover for this possibility.

I was asked to help in modifying a line where the material being treated in one section could not be allowed to wait more than seven minutes, otherwise it would harden and be useless.

The original designer had apparently assumed that nothing would ever break down, and a particularly optimistic forecast it proved to be. No facilities whatever had been provided for parking or dumping the over-aged material. Any delay over seven minutes was automatically an hour and a half. Men had to clamber into inaccessible spots and shovel or suck out with vacuum hoses the dust-provoking and now useless granules, which covered everything in sight with fine grit and so encouraged further breakdowns, especially in the electrical equipment.

The whole layout had been designed to fit into an existing building rather too small for it and so the flow of material had to follow a vertical zigzag route. There was no room to park or dump anything. Here, again, the mistake had been in the initial design. When thinking out a production line remember that it will inevitably be a non-production line sometimes.

If your process permits it, much the most economical way of dealing with stoppages is to return the material back to an earlier stage where it can join the waiting queue. This is standard practice in a multitude of different factories, and on the possibility of doing it may depend the profitability of the whole project. It also has the advantage that repairs and adjustments can be done when they are needed and do not have to be left to the weekend. You can combine quality and quantity more easily.

But this facility for sending anything faulty back to the start again can be abused and needs a quick-acting cost system as a safeguard. Quality may be obtained at the expense of quantity.

I knew a factory where there was no immediate direct check on the feedback of material and thus no precise explanation for the daily variation in labour costs for a unit of production.

After observing a typical period, I worked out statistically that an appreciable weight of material must have been having free rides round the factory without ever coming out of it for the whole of the previous ten years.

But perhaps the greatest importance of the discipline of logic lies in its protection against wishful or superficial thinking.

A logically deduced design must have something from which it can be deduced. The initial and necessary data must be there. And sometimes they are not. This means we must assume values for the missing facts, otherwise we can make no logical progress. The vital point is how far our subsequent conclusions are due to our original home-made assumptions. The ability to recognize that conclusions are only disguised assumptions is best based on detached reasoning. And this is easier to say than do. The disguises are endless. Sunday's joint may turn up on Monday as an unrecognizable wreck of cold meat and be interned on Tuesday in shepherd's pie. It is still the same meat.

An added difficulty is that designers have necessarily learnt to accept, without question, a large number of assumptions which they have never, in fact, had a chance of verifying. We unthinkingly and automatically use in calculations the acceleration due to gravity and fifty other essential constants. We must do. Unfortunately, we may unthinkingly accept some less well documented.

But this is not all. We must also accept and use, on occasions, assumptions which we know to be wrong, or at least to be only approximations, as otherwise our calculations would become bogged down in too many variables.

Probably the first example of a simplified assumption that we have met was at school when we were told that friction was dependent on pressure but not the area of the contacting surfaces. I remember receiving this news with the feelings, and doubtless the appearance, of profound scepticism. I proceeded to argue, in what must have been an extraordinarily irritating way, that a coasting 3-wheeled car would always skid equally front and rear, if this law was true. As, in practice, such a vehicle skidded at the rear, I could accept the law that centrifugal force acted at the centre of gravity, or the proposed law of friction, but not both at once. These remarks were not well received. On occasions we all have to use approximations and it is safe to do so provided we recognize their influence on the final conclusion. After all, 'absolute accuracy' is as much a fairyland phrase as 'instantaneous' or 'simultaneous'.

The real danger lies in slipping in a wrong assumption, or even a doubtful one, without realizing how profoundly it alters the subsequent logical verdict. By and large, the possibility of being deceived by an assumption masquerading as a conclusion is the designer's greatest enemy.

A simple example would be to consider a hook supporting a weight (Fig. 31). The hook is fastened by its shank passing through a plate and a nut on the top side. The nut is then tightened until the tension in the shank reaches the figure of 100 lb. Next a 100 lb weight is hung on the hook. What, we are now asked, is the resulting tension in the shank? We reply that we cannot calculate this without knowing the resilience of the materials. We are told that the plate and the hook are made from different materials, one is rigid, the other elastic. We are not told which is which. We reply that if the plate is resilient and the shank is not, we calculate the tension will be 200 lb. If the reverse is true and the plate is rigid and the shank elastic, the tension will remain at 100 lb.

Thus, if we say, 'The calculated tension is 200 lb', what we really mean is, 'My guess is that the shank is rigid.' Similarly,

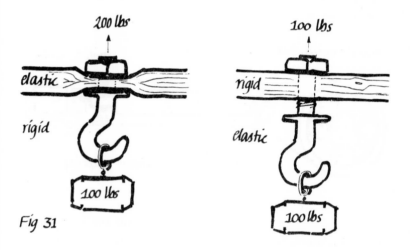

Fig 31

if we give the tension as 100 lb we are only saying that we think the plate is the rigid one. The word 'calculate' merely gives an undeserved impression of authority to a disguised guess. And the better the disguise the greater the danger.

Sometimes we tend to discount the guesswork element in our own picture thinking of how a mechanism actually works, while being very critical of other people's subjective ideas.

Consider the design of a portable machine for the vibration of concrete. There are three main requirements for a device of this kind. The first is that it must have a useful effective range, which means the vibrations produced must be of high frequency and acceleration. The second is that the diameter of its containing tube must be small and so displace as little concrete as possible. Finally, it must be portable.

The use of compressed air as the power medium is a good solution except for the fact that its efficiency is low. Electricity is more efficient but limits the output rev/min to 3,000 and thus puts a ceiling to the possible frequency of the out-of-balance weight which is the normal method of producing the vibrations.

The mechanism which we will now assess is designed to obtain from a 3,000 rev/min input a frequency and acceleration of vibration much greater than that which the out-of-balance weight can give.

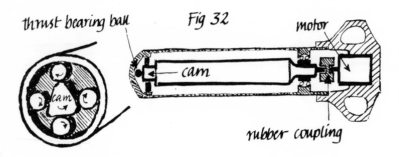

The source of the vibrations lies in a steel rod, driven at one end by an electric motor and a universal joint, and supported by a self-aligning bearing. The other end terminates in a triangular cam rotating between four equally spaced rollers. As the circumscribing diameter of the cam is greater than the distance between the faces of opposing rollers the rod is displaced from its central position each time a cam tip passes a roller. The result is that there are twelve deflections of the cam each time it revolves once relative to the rollers, which are located in a cage. The question that now arises concerns the frequencies of the vibrations produced when the cam-ended rod is rotated at 3,000 rev/min. The answer you give is dependent on the mental picture you form of how the cam and rollers will behave. (Fig. 32).

Someone might say, 'It is obvious that the behaviour will be dominated by the huge radial forces resulting from the combined centrifugal and impact forces. It is therefore safe to assume that no slip takes place circumferentially. The operating frequency can therefore be calculated directly from the dimensions and the geometry. It is an epicyclic gear that employs friction instead of gear teeth.'

Someone else may say, 'It is obvious that the behaviour will be dominated by the energy considerations, especially those resulting from the damping effect of the surrounding concrete. The energy produced and absorbed is dependent on the torque reaction of the roller assembly acting against the cam. This in turn is dependent on the total maximum friction available between the rollers and their outside race. In short, immediately the input torque from the driving motor begins to exceed the torque due to friction on the rollers, the latter will speed up their orbiting until the balance is restored. Therefore the frequency of vibration is dependent on the damping of the concrete.'

A third person may come along and say, 'It is obvious that the behaviour will be dominated by the huge accelerations due to both centrifugal and impact forces. The accelerations resulting from the movement of the cam and the resulting shock loads will greatly exceed the acceleration of the rollers due to their rotation around the race. This means that the rollers will bounce up and down against the race. Any attempt to restrain them by the cage would cause rapid wear and energy loss through friction, and involve pre-loading.

'The working principle is therefore one of churning the rollers round in front of the cam as they bounce about indeterminately.'

You can go on making up mental pictures like these for quite a long time, but you must remember that any frequencies that you calculate from them are little more than identifying labels for these pictures. They say nothing about whether a picture is of an actual or wholly imaginary scene.

Now we must turn to quite another way in which a disguised assumption can insinuate itself into our conclusions.

I discovered this early in my work as a consulting engineer. I will try to reproduce my reasoning on that occasion and I wonder if the erroneous assumption will be very obvious or not.

I was engaged in carrying out research into the behaviour of reinforced concrete over a wide range of conditions. Our research laboratory soon learnt to make concrete whose strength could be accurately predicted because we had carefully measured and controlled the proportions of the materials, especially that of the water–cement ratio, which was the crucial factor.

Our precision of control, I found, contrasted sharply with the methods and results current in those days on actual construction sites. All this happened about thirty years ago, and today's sophisticated methods of materials and mixing control were in their infancy.

The usual criterion for the suitability of a wet concrete mix was called the 'slump test'. This measured the degree of collapse of a concrete cone after the withdrawal of a metal dunce's cap in which it had been initially compacted. The result of this test was dependent on several factors, of which the compacting method used was one, but as an indication of the water–cement ratio it seemed indirect and unreliable. To find a direct form of measurement was an obvious need.

The problem was complicated by the fact that the two elements, cement and water, had to be compared in a mixture of aggregates and sand. Although having an almost irrelevant influence on the final compressive strength, the aggregates and sand were the predominating elements measured by volume. The vital mixture of water and cement hardly more than wetted their surfaces.

It therefore seemed essential, as a first step, to be able to segregate out the water and cement. A clue about how this might be partially achieved was provided by watching a concrete vibrator at work. If the vibration was continued after the maximum compaction of the concrete had been reached, a segregation effect began to take place. A milky pool of a sand-water-cement combination began to form around the tube of the internal vibrator. The concrete began partially unmixing itself. A sample taken from this liquid surrounding the vibrator showed that it was primarily a combination of water and cement with fine sand added.

Various methods of testing were then tried, and it was found that the electrical resistance to a direct current was dependent on the water–cement ratio, provided that the type of sand did not vary too markedly. Here then was the basis of a testing machine and I proceeded to exploit it by designing and patenting the instrument shown (Fig. 33).

This consisted in principle of a small cylindrical box with a battery operated vibrator built into it and two electrodes on the under side. The whole was mounted at the end of a walking-stick-type handle and there was a dial showing the electrical

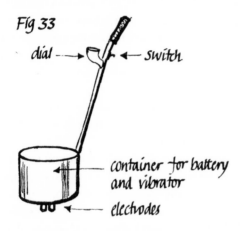

Fig 33

dial ---→   ←— switch

container for battery and vibrator

electrodes

resistance between the electrodes translated into equivalent water–cement ratios.

To test wet concrete on site all that was necessary was the insertion of the electrodes into the mix. This could be done before or after the concrete had been laid. A switch was then operated and the electrodes vibrated violently for a few seconds until a small volume of water-cement-sand mix was formed between them. That this condition had been achieved was shown by the current reading becoming stationary, and the water–cement ratio could then be read off.

Unfortunately somewhere in the foregoing there is a guessed assumption treated as a logical conclusion.

It is that I assumed that anyone would want to buy such an

instrument. The word 'obvious' in the phrase 'an obvious need' should have given you the clue. Always be suspicious of the 'obvious'. It is a word that is often used to camouflage the illogical. It attempts to bulldoze argument.

In those days market research was not as readily available as it is today, but the danger, in principle, still remains. The danger of assuming that what people need is necessarily what people want to buy.

Hardly anyone wanted an instrument like that on a contractor's site. The man on the mixer who had been controlling by sight and slump did not want to learn new tricks. His supervisor did not want to introduce a means whereby his boss could make random checks on what was happening. The efficient contractors had control and inspection systems already established. Some did not want to know too much, others thought they knew it all.

A designer must face up to economic realities. The customer may be wrong but you must often treat him as if he were right. It's his money you are after.

There is also another way, even more unpleasant, in which financial considerations intrude into the designer's world. A completed, or partially completed design, embodying many months of conscientious work, may have to be abruptly cancelled, and the whole project abandoned due to changes in the economic climate. Excellent although the design may be from the engineering point of view, it is useless to continue if its basic principle has become out of date.

Sometimes boards of directors do not sufficiently appreciate the disappointment and lowering in staff morale that such cancellations may cause. Financiers may decide to cut their losses and sell out unprofitable shares and re-invest in something more promising and only ledger entries are affected.

Cancel an engineer's project and you cannot re-invest all the care and thought that has already gone into it. That has gone for ever; an active part of his creative life, almost a part of his life itself, will be buried for ever in some dusty drawer. Often it can't be helped, but the reasons for the decision should be sympathetically outlined and discussed with the designer. You have to be one to realize how heartbreaking it may be.

There is a third way in which the economic may exert unwelcome pressure, and it is one that should be resisted as strenuously as possible. Success in combating it may well depend on the designer's position in the organization. He may have little option but to do what he is told.

A group of companies who were very involved in the mixing of materials were approached by an organization sponsoring a new type of machine for this purpose. I was asked to give a report on its performance.

The demonstration of a prototype machine was arranged in a Midland factory and the raw materials were sent up in advance. These consisted of oxidized oil, physically like a sticky and soft rubber, wood flour and finely ground whiting. This trio had to be homogeneously mixed together. It was most difficult to do.

On arrival at the factory I was introduced to the inventor, his machine and his sponsors. It appeared to me that the machine was essentially a disintegrator, not a mixer, and the inventor immediately agreed. However, he added, the simultaneous disintegration of materials might mix them up together. I had the feeling that the inventor was under considerable pressure from the commercial side to widen the market for his machine by using it in this way, but I may have been wrong; but it was quite clear that the machine had never been tried out as a mixer.

We all gathered round and the demonstration began. Some hundredweights of each of the three raw materials were dropped into the hopper above the machine and we waited expectantly at the outlet orifice to examine the result. Unfortunately nothing at all came out. The machined hummed happily to itself and that was all that happened. After about ten minutes the inventor said he thought something had clogged up inside and he felt that this could be rectified by pouring in a few gallons of raw linseed oil. This was done. The machine hummed more contentedly than ever; nothing else happened.

After an embarrassing pause, the inventor remarked that the machine needed something to 'clear it out' and departed rapidly in the direction of the car park, saying over his shoulder that he would be back in a minute.

A quarter of an hour later he reappeared carrying a sack on his shoulder, climbed up the ladder and poured its entire contents into the hopper. We saw with some alarm that this consisted of wood shavings. The machine gave a surprised yelp and the ammeter indicator on the motor flew off into the red on the dial, but nothing else happened. Soon the machine was back to humming again. The inventor looked worried and hustled away once more. This time he returned with two sacks of household coke which he poured rapidly into the hopper. Instantly bedlam broke out. The machine screeched, the floor shook, and the ammeter flew off the dial, but nothing came out. The appalling noise continued and the outside of the casing began to heat up rapidly. I was not certain about the physical characteristics of a combination of oxidized oil, raw oil, whiting, wood shavings and coke, but I imagined that once they reached a critical temperature something violent would happen.

The inventor rushed away again and I thought that I should take a brief walk outside the factory until he returned. When he did so, I had to help him lift out of the boot of his car a small but very heavy sack. 'What's in this?' I asked. 'Lead shot', he said. We dragged the sack into the factory and our combined efforts hoisted it up to the hopper. In went many thousands of small lead pellets as used in shotgun cartridges. For five seconds the noise defied description; then, without warning, the machine split vertically in half, and we were retreating down the factory pursued by a boiling black polyglot flood.

We heard nothing more about the machine, but my sympathy is with the inventor. The principle that he had used is now standard practice in many types of disintegrating machines all over the world. He had wrecked everything by taking an uncalculated risk. Sometimes we have to take risks, it is part of the adventure of engineering, but let them be calculated ones. Don't dismiss logic and rely on luck. Look hard before you leap and this means looking for a logical way round that would make leaping unnecessary. If you must leap, don't do it in public.

Although positive in purpose this chapter has been largely negative in content. This has been done on purpose. To read about how clever other people are, I find rather depressing; disasters are more entertaining and therefore more memorable.

However, a positive summary of principles can be listed as follows:

(1) Think logically
(2) Remember money is valuable
(3) Avoid technical snobbery
(4) Design for a predictable life
(5) Arrange for quick replacement of wearing parts
(6) Protect with safety devices
(7) Park, dump or return
(8) Watch for disguised assumptions
(9) Beware of unwise commercial pressure

6

SAFETY MARGINS

We all automatically insure against risks, often very remote ones. Everyone takes out a policy against fire destroying a new house, which fortunately is a rare occurrence.

By contrast the insurance of a new design against much less remote disasters is not so automatic. Temperature stresses, torsional vibration, feedback oscillations and similar destroying forces may break out without warning with devastating results. Nearly all machines have some potential disaster lurking in the background ready to pounce. The best protection is a large margin of safety.

In the last chapter we said something about foreseeing likely difficulties in keeping a machine in production and how to guard against them. Here we are concerned with the unlikely ones. We would do well to work out how unlikely they really are and then, if necessary, insure against their happening at all.

I remember being called to examine the largest and most expensive chilled iron roll ever made in Europe. It had just fallen in half. The astounded operators explained that there had been no exterior load on it at the time. 'It just broke itself', they said. And they were right.

The wall of the roll was drilled for accurate temperature control by hot water circulation. The heat exchanger through which the water passed was large and efficient; too large and efficient. The rate of change of temperature in the water could, if unwisely operated, become many times too high to avoid dangerous temperature stesses in the roll. If the designer had recognized the possibility of this happening in theory, and insured against it by limiting the size of the heat exchanger so that it was physically impossible to produce a too rapid rate of change of temperature, all would have been well, whatever the operators did. Instead he left it to luck and did not have any.

Another example was where an engineering firm in this

country was constructing a production machine from detail drawings sent over from the U.S.A. The moving masses were driven from one end, and conditions for torsional vibration were theoretically possible but thought very unlikely. The machine thought otherwise: the torsional vibration was terrifying to behold; the rear end seemed almost to be revolving backwards some of the time. Economic reasons dictated that production should start up and we had to design and rush through a drive putting power in each end. But the torsional vibration beat us and before the new parts were ready a 10 in. diameter steel forging had sheared in half. It was suicide, not murder.

But the mistake which I remember most vividly was the one I made myself.

The special car that we last left embedded in the Somerset soil was dragged out, cleaned and repaired. Its performance suggested that with some extra streamlining it should be capable of capturing the standing-start kilometre and mile records for its class. I decided to go to Brooklands track and do some timed runs, but first I must alter the gear ratios. A quotation from a gear-cutting firm for providing a new final drive instantly eliminated, for financial reasons, any attempt to do it this way. I must find something cheaper.

The front twin-tyred wheels needed replacement, as track racing tyres could not be had in that size and in any case they made the steering rather peculiar at three-figure speeds.

As new front wheels would have to be paid for anyway, why not make them 40 per cent larger in diameter, at little extra cost, and so achieve a higher effective gear ratio too? This I decided to do, being more concerned with avoiding stresses in Barclays Bank than anywhere else.

The wheels were built and new track tyres fitted and off we went for some practice runs. It was a problem to know what to do with the stopwatch, for I wanted to time myself and so know how I was doing without delay. I finally found it best to hold the watch in my left hand, although this made gripping the steering wheel awkward. The first two runs indicated that the records should be within our capabilities and I decided to try hard, using crash gear-changing.

The start went well and soon the rev counter indicator was well into the red section of its dial in top gear. Then, without warning, both driving wheels locked completely solid. I hastily grabbed the steering wheel with both hands, dropping the stop-watch into the gear-change linkage, where its back flew off. It was never quite the same again.

Meanwhile things were happening fast. I soon ran out of steering, trying to deal with skids; next the scenery seemed to start spinning like a top. At long last we stopped. I crawled out and examined the state of the car. The front wheels had clamped solid at over 100 mi./h and remained stationary; broken gear teeth had jammed up everything. On each driving tyre the rubber had been planed and melted off in a dead straight line, right down to the inner tube.

I had overlooked the fact that the stresses in a transmission are not primarily due to torque of the engine but the torque reaction of the wheels. Although the engine power was the same, the 40 per cent increase in wheel diameter had increased the forces in the transmission correspondingly, and I had never checked up to see how near the limit I was. Now I was a poorer and wiser man. I returned the tyres to Dunlops, complaining that 'they had worn out rather quickly'. When they had stopped laughing they gave me another pair free of charge, which was most kind of them and more than I deserved.

Circumstances are not likely to be so kind to a designer who relegates the improbable to the impossible and forgets about it. Always check up.

One of the great values of academic knowledge is that you can generally assess how near you are to dangerous stresses or instability of some kind, and this is especially useful when the complexity of the system defeats picture thinking.

An example of this can be found in the weighing control system which we have already discussed and shown in Fig. 20.

I asked a Cambridge colleague of mine to read the earlier chapters of this book as soon as I wrote them. He, as an expert on control theory, was deeply shocked that I had not given any warning about blindly copying the principle of Fig. 20 (p. 49).

'Your semi-static analysis solution', he wisely said, 'proved satisfactory, but in slightly different circumstances might have led into awful trouble. The difficulty is that at each step a small phase shift occurs, and if this happens to tot up in total to 180° (another 180° is cancelled by the negative sign inherent in your feedback loop), you may be in trouble. Actually things are potentially worse than this, because even your analysis shows that you must lose about 90° due to the inertia of the motor/belt drive, and therefore you only have 90° with which to play elsewhere.'

Whether this margin is dangerously small can only be established by analysing the system.*

So far we have assumed that there is no difficulty in distinguishing between the main and marginal problems, but in practice it is possible to confuse them. Sometimes the difficulty which appears remote is actually the major one.

Suspension bridge designers thought at first that the major problem would be the load, and learned by spectacular experience that it was really the stability.

During the Second World War special concrete was required to replace armour-piercing steel, and a specification was evolved so that the energy could be transferred from the shell to the concrete very effectively. The shell stopped in a few inches. Unfortunately a lump of concrete, travelling nearly as fast as the shell and even more uncomfortably shaped, would fly off the back of the slab, and act as a most lethal projectile. The basic problem was to contain and not transfer energy.

Another wartime example was in the Mulberry Harbour. Large breakwaters made of concrete had to be so constructed that they could be towed across the Channel and then sunk at the right spot on the Normandy beach.

Fig. 34 shows the way they were designed. It will be seen that, as long flat-bottom barges with a central division, they could

* See D. B. Welbourn, *The Essentials of Control Theory for Mechanical Engineers* (Edward Arnold, 1963).

float and then be rapidly sunk by operating the two rows of valves in each side.

What was wrong with this design? Of course, once you know that there is something wrong, you can settle down and discover what it is. In this case, it is that when the valves are opened the water will not enter in equal volumes through all of them as the breakwater will never be exactly vertical even in a calm sea. Immediately one side begins filling up faster than the other, the breakwater will tip towards that side and make matters worse.

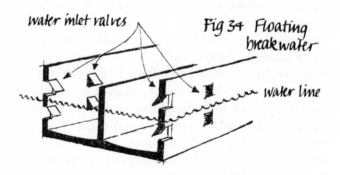

Soon the ports on the lighter side will be lifted out of the water and the breakwater will sink on its side and not its bottom.

Tests were carried out in time for modifications to be made to the final design and all was well.

The point is that there is no one normally there to warn you that there is something wrong somewhere and so you must develop a critical attitude to your own work. This is not easy and introduces us to a point of major importance.

Throughout this book we have talked about 'designers' and 'engineers' and sought, by analysing their work, to find general and helpful maxims. But there is a danger in doing this, for we may forget that a 'designer' or 'engineer' is first and foremost a human being. A human being who designs or engineers has, in addition, a wide range of complicated and interacting emotions. It is easy to become emotionally involved in a new invention and so resent criticism of it even when that criticism, if heeded, would turn a failure into a success. Confidence may be blinded by conceit or undermined by an inferiority complex.

History has shown that many new inventions were not inventions at all; they were re-inventions. The idea was not new but the determination to press through with it was.

Character is the foundation of ability. Develop as large a safety margin as you can against such things as laziness, impatience, prejudice and arrogance and all the other human failings that inhibit constructive achievement in all spheres, especially that of creative engineering design.